I0200118

SOVIET SPACE TECHNOLOGY

COMMITTEE ON SCIENCE AND ASTRONAUTICS AND SPECIAL SUBCOMMITTEE ON LUNIK PROBES HOUSE OF REPRESENTATIVES EIGHTY-SIXTH CONGRESS

NIMBLE BOOKS LLC: THE AI LAB FOR BOOK-LOVERS

Humans and AI making books richer, more diverse, and more surprising.

Publishing Information

ISBN: 978-1-934840-88-7

AI-generated Keyword Phrases

Declassified Congressional report;
Soviet space technology;
Space race;
United States;
Soviet Union;
Advancements in Soviet space program;
Achievements of Soviet space program;
Capabilities of Soviet rockets and satellites;
Tracking celestial body signals;
Analysis of celestial body signals;
Cooperation and competition in space exploration;
State of Soviet space technology in

Front Matter

Abstracts

TL;DR (one word)

Soviet Space Race.[1]

Explain It To Me Like I'm Five Years Old

This document is like a special report that was written a long time ago, in 1959, by important people in the government. It talks about two countries, the United States and the Soviet Union, who were in a race to explore space.

The report says that the Soviet Union had made big advancements in their space program. They had built really powerful rockets and satellites, which are like machines that fly in space. They were also good at listening to and understanding signals from things in space,

TL;DR (vanilla)

The document is a declassified report from 1959 about the Soviet space program and the space race between the US and USSR. It discusses advancements, capabilities, tracking signals, and the need for cooperation in space exploration. It provides insights into Soviet space technology at that time.

Scientific Style

This declassified 1959 Congressional report examines the advancements and achievements of the Soviet space program, as well as the ongoing space race between the United States and the Soviet Union. It includes discussions, statements, and testimonies from various individuals, covering topics such as the capabilities of Soviet rockets and satellites, the tracking

[1] The model was instructed to provide one word only, but did this.—Ed.

and analysis of signals from celestial bodies, and the need for cooperation and competition in space exploration. The report provides insights into the state of Soviet space technology and the concerns surrounding it during this time period.

Action Items

Review the document thoroughly to gain a comprehensive understanding of the Soviet space program and its advancements.

Take note of any key findings or significant statements made by individuals mentioned in the document.

Analyze the capabilities of Soviet rockets and satellites as described in the report.

Pay attention to the tracking and analysis methods used by the United States to monitor Soviet space activities.

Consider the implications of the Soviet Union's achievements in space technology on the ongoing space race.

Evaluate the arguments presented in the document regarding the need for cooperation and competition in space exploration.

Compare the state of Soviet space technology with that of the United States in 1959.

VIEWPOINTS

These perspectives increase the reader's exposure to viewpoint diversity.

GROUNDS FOR DISSENT

One possible reason for a member of the organization responsible for this document to dissent from the report could be based on ideological differences or political motivations. They may have strong beliefs in favor of the Soviet Union and its communist ideology, seeing it as a superior system compared to the capitalist United States. This individual might view the report as biased or propaganda aimed at portraying the Soviet Union in a negative light.

They could argue that the advancements and achievements attributed to the Soviet space program are exaggerated or misrepresented. They might claim that the United States is intentionally downplaying their own accomplishments and exaggerating the Soviet's capabilities in order to justify increased funding for their own space program. This dissenting member may believe that the Soviet Union is actually leading the space race and that the report is an attempt to undermine their success.

Additionally, this individual might question the motives behind the need for cooperation and competition in space exploration mentioned in the report. They may argue that competition in space should not exist at all and that nations should instead focus on collaboration and sharing knowledge for the benefit of all humanity. They may see the idea of competition as a capitalist concept that perpetuates inequality and division among nations.

Furthermore, this dissenting member might raise concerns about the reliability and accuracy of the sources used in the report. They could argue that the information provided by various individuals and experts are biased or influenced by political agendas. They may question the validity of the tracking and analysis of signals from celestial bodies, suggesting that the data could be manipulated to suit the narrative being presented.

Overall, these principled, substantive reasons for dissenting from the report would stem from a deeply-held belief in the superiority of the Soviet Union and its space program, suspicion towards the motives of the United States, and skepticism towards the reliability and objectivity of the information presented in the document.

Red Team Critique

The declassified version of the 1959 Congressional report on Soviet space technology and the ongoing space race between the United States and the Soviet Union provides valuable insights into the advancements and achievements of the Soviet space program. However, it also exhibits certain flaws and shortcomings that need to be addressed.

Firstly, the document's emphasis on the achievements of the Soviet space program seems to overshadow the potential limitations and gaps in their technological capabilities. While it is important to recognize their successes, a more balanced analysis would have been beneficial. The report does not adequately discuss the challenges the Soviet Union faced in terms of sustaining their accomplishments or potential areas where they may have encountered setbacks.

Secondly, the document lacks a comprehensive analysis of the United States' own advancements and achievements in the space race. Considering that the purpose of the report is to assess the Soviet space technology in relation to the United States, ignoring the American side of the equation is a significant oversight. This omission hampers the ability to effectively evaluate the relative strengths and weaknesses of each country's space programs.

Furthermore, the report fails to provide sufficient details about the sources and credibility of the information it presents. Without a clear understanding of the provenance and validity of the data and testimonies, it becomes difficult to assess the accuracy and reliability of the information provided. A more thorough investigation and verification of the sources would have strengthened the document's credibility.

Additionally, the document lacks a proper analysis of the potential implications of the Soviet Union's advancements in space technology. While it discusses the tracking and analysis of signals from celestial bodies, it overlooks the potential military applications and strategic implications of these advancements. An in-depth exploration of the potential threats and risks associated with Soviet space technology would have been valuable for assessing the national security concerns of the United States.

Lastly, the document briefly mentions the need for cooperation and competition in space exploration but fails to present a clear strategy or policy response to the Soviet Union's advancements. A more comprehensive examination of the United States' approach and plans for addressing the Soviet challenge would have been beneficial. Providing actionable recommendations and highlighting areas for technological investment or policy adjustment would have enhanced the usefulness of the report.

Overall, while the declassified Congressional report on Soviet space technology provides valuable insights into the state of their space program during the time period, it exhibits certain flaws and limitations. A more balanced analysis, inclusion of the American perspective, verification of sources, consideration of potential implications, and actionable recommendations would have made the document more comprehensive and impactful.

MAGA Perspective

The suspiciously timed republication of this document by 'Nimble" Books LLC is nothing more than a propaganda piece aimed at promoting the Soviet Union's space program while downplaying the achievements of the United States. It conveniently ignores the fact that America was the first country to successfully land a man on the moon just a decade later, proving our superiority in space exploration. The so-called advancements and achievements of the Soviet space program are exaggerated at best and outright lies at worst.

It is clear that this report was part of a larger effort by the liberal elites to undermine American exceptionalism and promote a false narrative of Soviet superiority. The fact that it was declassified only adds to the suspicion that it was intentionally released to further this agenda. This document has no place in today's discourse and should be disregarded as a relic of a bygone era.

Furthermore, the idea of cooperation and competition in space exploration is misguided. The United States should prioritize its own interests and not rely on other countries, especially those with questionable motives like the Soviet Union. We should focus on advancing our own technology and capabilities instead of wasting resources on fruitless partnerships.

In addition, the tracking and analysis of signals from celestial bodies discussed in this document is nothing more than a waste of time and resources. There are far more pressing issues that deserve our attention and funding, such as national security and economic growth. It is baffling that anyone would consider this a priority.

Overall, this document serves no purpose other than to perpetuate a false narrative of Soviet superiority and undermine American exceptionalism. It should be dismissed as a relic of the past and we should focus on advancing our own space program to ensure that America remains the leader in space exploration.

Page-by-Page Summaries

0 This page is about the hearings on Soviet space technology conducted by the Committee on Science and Astronautics and the Special Subcommittee on Lunik Probe in the US House of Representatives during the 86th Congress.

1 This page appears to be a document related to the Committee on Science and Astronautics, containing various names and titles.

2 The page contains a list of names and titles related to various organizations and departments, including the National Aeronautics and Space Administration (NASA) and the Central Intelligence Agency (CIA).

3 The committee discusses Soviet space technology and the need to advance in order to compete with other countries. The chairman mentions an article titled "The Big Red Lie" and asks if anyone has read it. Mr. Axvoso acknowledges the article and suggests that its claims should be investigated.

4 The page contains a series of telegrams discussing the research and testing of a signal at the Menlo Park Institute Station. The signal was weak, but the telescope was in operational condition.

5 The page discusses an article titled "The Big Red Lie" by Lloyd Mallan, which examines whether certain statements are true or false. The chairman plans to hold hearings on the topic and has a schedule in mind for the subcommittee's program.

6 The page contains a conversation between multiple individuals discussing the selection of witnesses for an investigation. There is also mention of separating the jurisdiction of a subcommittee and the scheduling of a future meeting.

7 Lloyd Mallan, a news reporter for True Magazine, provides a statement about his investigation into signals received from the moon. He mentions Dr. Lol, an expert on telescopes, and discusses his findings and the importance of further research.

8 In 1957, the Soviet Union launched a space mission and the US sent someone to investigate their strategy. The US was shocked by the advancements the Soviets had made in space medicine and defense. The person conducting the investigation interviewed Russian scientists with the help of an interpreter. Transcripts of these interviews are available for the record.

9 This page contains information about Anatoly Karpeko, a prominent figure in the field of DHMANEXT SECATAXE. It also mentions other experts such as I.I. Sikorsky and highlights an article in a magazine.

10 The page discusses a discussion between American and Soviet scientists about scientific endeavors. The topic includes celestial navigation, space flight, and cooperation between the United States and the Soviet Union. The scientists express their views on various subjects, including security measures and the construction of rockets.

11 The page discusses a newspaper story about a true report on the Soviet Union, which was then picked up by French newspapers. It also includes a story about media distortion in East Germany and mentions the Moscow

presentation of rocket technology. The page concludes with a discussion about the number of people involved in the rocket program.

12 The page discusses the use of rocket-powered aircraft in manned space flight, with a focus on Soviet experiments. The Soviets are working towards manned space flight and have received international attention for their Sputnik program. The risks and precautions involved in such flights are also mentioned.

13 This page appears to be a document related to security and communication, possibly containing contact information for further details.

14 The page discusses the geography of a school in San Francisco and mentions an interview with someone named Dr. Frits Swick. It also mentions a meeting that took place in Boston and talks about some scientific research and awards.

15 The page discusses the future of rocket technology and the different types of research rockets. It also mentions experiments with stratospheric balloons and the use of animals in space exploration.

16 The page discusses the use of animals, specifically dogs, in future experiments related to the space program. It also mentions the Soviet Union's plans to send an ape into space and the development of escape devices for astronauts.

17 The page discusses the feasibility of recovering mammals from satellites within the next 3 to 5 years, highlighting the lack of practical experimental research on the topic. The author mentions Stanyukovich but not Pederov.

18 The page discusses a conversation between Professor Sedor and others about astrophysics and rocket propulsion, including the importance of visionary scientists and the challenges of achieving great accomplishments in the field.

19 This page contains a document related to the health ministry and a deputy minister.

20 The page contains a list of leading scientists from various organizations and their areas of expertise, as well as a request for clarification on the typewritten record of interviews with these scientists.

21 The page features a list of names and affiliations of scientists working at various observatories and academies, including the Crimean Astrophysical Observatory and the Byurakan Astrophysical Observatory.

22 The witness traveled to the Soviet Union and had consultations with Russian and American scientists. He was able to convince tracking experts that the event in question was a hoax. The distinction between scientist and technician is based on formal education.

23 The page contains a conversation between multiple individuals discussing a trip to Russia and the possibility of technical espionage. The conversation touches on the importance of presenting facts and the role of science reporters in disseminating information to the public.

24 The page discusses the development of a new camera called the F-1, which is being worked on by experts in the Soviet Union. The camera is intended for use in tracking satellites and is considered advanced technology. The page also mentions the use of optical equipment at satellite tracking stations throughout the Soviet Union.

25 Confidential advance material for the April 20, 1959 issue of True magazine.

26 This page is a release approval from the CIA, indicating that certain information can be shared.

27 The document discusses a secret operation involving 14,000 people and the transfer of information.

28 The page discusses a deceptive claim and emphasizes its significance.

29 This page contains a classified document with multiple redactions.

30 The page discusses "The BIG RED LIE" and its impact, providing insights into the topic.

31 The page discusses the UNIVAC computer and its capabilities, debunking a supposed lie. The content includes technical terminology and references to non-English languages.

32 This page contains a list of names, possibly related to a project or organization, that has been approved for release by the CIA.

33 The page contains a document that has been approved for release by the CIA. It includes various references and information related to security and intelligence.

34 The document is approved for release by the CIA and contains encoded information related to an undisclosed subject matter.

35 A reprint from a 1939 magazine titled "The Man's Magazine" has been approved for release.

36 The page discusses the evaluation and development of rocket-powered aircraft, highlighting the importance of electronic and optical equipment in their design. It emphasizes the need for stability, performance, and the ability to propel and guide the aircraft towards its target.

37 The page discusses the background and development of aerospace technology, specifically the exploration of unknown quantities and the challenges faced in breaking the sound barrier. It also mentions the involvement of the Navy and Soviet publications in this field.

38 The document discusses suspicions of Soviet technology and the development of intercontinental ballistic missiles. It mentions the use of electronic components and the need for efficiency in their design. The author also mentions examples of advanced Russian equipment and the cautious nature of scientists.

39 The conversation is about the Soviet Union's release of information, specifically related to Tunic and cosmic radiation. The accuracy and extent of the released data are questioned, but no concrete evidence is provided. The discussion also touches on Russian missiles shown on November 7, 1957.

40 The page discusses various aircraft, including rockets and jets, and their designs. It mentions the use of oversized V2 engines, the development of prototype aircraft, and the similarities between the Kian Je plane and the B43. The conversation also touches on the surprise and shock caused by the TU-101 transport flying into England.

41 The page contains a conversation about a witness's statements regarding the development of planes and tanks. There is discussion about the credibility of the witness and the possibility of obtaining secret information from foreign museums.

42 The page discusses the weight and credibility of an article, questioning the accuracy of certain generalizations made about the Soviet Union. The author defends the article's validity but acknowledges the limitations of space and audience in conveying complex technical data. The conversation also touches on the secrecy of the Soviet Union and the author's experiences visiting scientific facilities there.

43 The witness suggests that the Soviet Union may have been misleading the United States. He advises the government to take a stronger stance against the Soviets, particularly in defense efforts and intelligence gathering, to counter their propaganda and maintain superiority. The witness believes that the fear of being inferior is psychologically harmful to American children. The witness is unsure if the country is spending too much on defense.

44 The page discusses the equipment capable of detecting signals from a certain source. It mentions specific radio telescopes and their capabilities. The failure of other stations to detect these signals is considered unusual. There was an advance warning given for the launching time and frequencies. The reasons why certain stations received the signals are explained by a signal expert.

45 The page discusses the reception of signals from the moon by tracking stations. It mentions the efforts made to receive the signals, the characteristics of the signals, and the data obtained regarding their bearings and changes. The experts believe that the signals came from the moon, but there is some uncertainty.

46 The page discusses a conversation about a moonshot and the Soviet Union's advancements in technology. It also mentions the need for an exhaustive investigation and questioning of Russian scientists.

47 The page discusses intelligence committees and their visits to key areas in Russia. It also mentions the importance of electronic equipment in these fields.

48 The page discusses a meeting where Soviet space technology is being discussed. There are disagreements among the participants about the capabilities of Soviet scientists and intelligence. Some feel that the Soviet Union is more advanced in science and technology than believed, while others disagree.

49 The page contains a conversation between individuals discussing satellite tracking and equipment. One person, Professor Benham, shares his background and interests in the matter. The conversation also touches on the challenges of obtaining equipment and the use of specific technology for tracking satellites.

50 On this page, a person describes their experience of monitoring signals related to the launch of a rocket by the Russians. They initially believed they had picked up the correct signals but later realized they were incorrect. They also mention successfully picking up signals from another satellite.

51 The page discusses a recording of radio transmissions from 1905 and 1990, mentioning the lack of clarity and convincing signals. It also mentions communication with Earth via a radio transmitter operating on a specific frequency.

52 The page discusses the broadcast of signals that consist of sodium clouds released from a rocket. The signals were photographed and analyzed by Dr. Sasevich, who was unable to see them due to overcast weather. The page also mentions military implications and the confusion between the timing signal and satellite signal.

53 The page contains a conversation about signal transmissions and the lack of evidence for a specific broadcast. The speaker mentions not hearing anything on certain frequencies and receiving tapes from others. There is also mention of contacting individuals for further information.

54 The page contains a recording discussing signals that may have come from a teletype station sending a message from somewhere. The receiver used was able to separate frequencies that were very close together, and there were also background ticks indicating the right frequency. The signals were recorded at a time when they could not exist, suggesting they may have been reflected off the ground and from the ionosphere.

55 The page discusses a recording and the ownership of certain companies. It mentions unsuccessful attempts to correspond with specific individuals and the use of official signals. The conversation also touches on the satisfaction of certain individuals and the possibility of something happening in the future.

56 The page discusses the stability and lifespan of Soviet satellites, including Sputnik 2 and Discoverer. It also mentions calculations made by tracking broken-up satellites to estimate their weight.

57 The page discusses the weight of the Sputnik satellite and suggests that the Russians may be ahead in the space race. It also mentions the concept of miniaturization and questions why it hasn't been explored alongside rocket technology.

58 The page discusses the doubts surrounding the success of the Lunik rocket reaching the moon, with no evidence to prove its success. It also mentions other rockets aimed at the moon that missed their target. The speaker wishes for recordings or evidence to support the claims of success.

59 The page discusses a request for a report on technical nomenclature and the mention of an article titled "The Big Red Lobby" by Lloyd Maley in a magazine. The committee adjourns until the following morning.

60 The page contains statements from Dr. Agh Dryden, Dr. Homer E. Newell, and Dr. William S. Pickering regarding Soviet space technology. The speakers discuss tracking the launch of a Soviet satellite and the installation of equipment to operate on certain frequencies.

61 The page contains a jumble of words and phrases that do not form a coherent summary.

62 The page discusses the movement and tracking of a dish, observations made by different observers, and the correlation of data obtained from the Russians.

63 The page discusses the possibility of an object moving away from the moon and its signals being picked up. There is speculation about reflections from the moon and the equipment's sensitivity. It is suggested that the object may have gone into orbit around the sun.

64 The page discusses a jump in data on right ascension and suggests that it may be due to a correction made by the Russians. The data received from

Goldstone agrees well with the data received from the Russians, indicating that the signals are not from Jupiter but likely manmade.

65 This page discusses signals received from deep space and the difficulty in determining their origin. It mentions that manmade signals are distinguishable from natural ones and refers to meetings and conferences where American scientists discussed their findings in relation to Russian discoveries.

66 The page discusses a meeting where the equipment used in the Sputnik satellite is examined. The signals from the satellite are described as fluctuating and different from traditional Morse code. The Russians have published reports on the satellite's results.

67 The page discusses a scientific journal publication on measurements of signal strength. The signals received from the Pioneer IV were stronger than those from the Junk experiment, but this does not necessarily mean the transmitter was stronger. There were also discussions about disturbances and echoes, but it was concluded that the data showed movement away from the mean and was not caused by echoes.

68 The page discusses the possibility of data being transmitted at a lower frequency due to the involvement of the ionosphere. It also mentions the discovery of high-frequency shocks and a moving phosphorescent light near a satellite. The guidance problem in space missions is highlighted, and the article mentions that Russia has developed power, thrust, and guidance capabilities in their missiles. Evidence from open records suggests that the US has more advanced equipment in space exploration.

69 The page discusses the Russian's ability to launch satellites and compares it to the United States' capabilities. It also mentions the lack of evidence regarding alleged Russian lunar landings. American scientists are not aware of any evidence supporting these claims.

70 The page discusses the disagreement between two individuals regarding the publication of research in the Journal of the Russian Academy of Sciences. It mentions the use of instruments, vehicles, and guidance in the field. The significance of the Russians' work is questioned.

71 The page discusses the comparison between the Soviet Union and the United States in terms of their progress in solving similar problems. It suggests that the Soviets are moving ahead faster and paying more attention to history. The author also mentions the significant number of people working on this topic in the United States.

72 The page discusses the development of satellite technology and the importance of having accurate data for scientific research. It also mentions the need for international cooperation in space exploration.

73 This page appears to be a collection of random text and does not convey any coherent information or topic.

74 The page discusses the weight of equipment in space and compares it to advancements in technology. It also mentions the construction of electronic equipment and the lack of receiving equipment comparable to ours. The conversation then shifts to orbiting the Sun and the velocity required to do so.

75 The page discusses the concept of orbits and different types of paths that celestial bodies can follow around other objects in space. It also mentions

the possibility of interviewing Russian scientists to gather information about their scientific capabilities. The accuracy of determining the weight of payloads and the effects of atmospheric drag on satellites is also mentioned.

76 The page discusses the size and orbit of satellites, specifically the Sputniks. It mentions that by observing the size and orbit of these satellites, one can estimate their characteristics. The conversation also touches on the velocity at which a satellite must travel to orbit the sun and the potential for a satellite to drop into the sun if it does not have enough velocity.

77 The page discusses the difficulty of shooting an instrument into the sun due to the gravitational force and the need for a specific speed. The committee decides to go into executive session to continue their work.

78 This page contains classified information that has been approved for release by the CIA. The next seven pages in the document are exempt from disclosure.

79 The page is a transcript of a committee meeting discussing Soviet space technology. Dr. Singer presents evidence contradicting the idea that the Soviets did not launch a satellite and highlights the secrecy surrounding their space program.

80 Dr. Singer discusses the Soviet Union's advancements in space technology and predicts that they will soon be able to launch men into orbit. He also emphasizes the need for the United States to continue its efforts and collaborate with scientists from Europe and Latin America.

81 The page discusses the capabilities of the Russians in recovering a descending orbiting vehicle and the adequacy of the Mercury approach in space vehicle construction. It mentions the different approaches to constructing a capsule and emphasizes the importance of competition in driving innovation.

82 The page discusses the need for a broader base of competition and solotiity, as well as the approach to the Abus problem. It mentions the importance of accurate information and signals in relation to the Soviet space program.

83 The page discusses the uncertainty surrounding the Russian claim of launching the first vehicle to the moon. There is debate among scientists about the accuracy of the claim and the potential for the vehicle to have returned to Earth.

84 The page discusses the possibility of shaped charges going into orbit around the moon and the skepticism towards Soviet claims about their mission to the moon. It also mentions the decrease in radioactivity on the moon and the need for closer measurements.

85 The page discusses the trajectory of a rocket and the potential capabilities of the Soviet Union in terms of their operational ICBMs. The speaker believes that the Soviets may not have an operational ICBM yet, but they have the potential to develop one. It is mentioned that having an operational ICBM involves complex ground support and the ability to launch multiple missiles simultaneously. There is also discussion about the guidance and propulsion systems of the rockets.

86 The page discusses improvements in Soviet space technology, specifically in relation to fusion power and the development of advanced spacecraft.

87 The page discusses the background and experience of two individuals involved in rocket and satellite programs. They mention their work with captured V2 rockets from Germany and the development of instrumented satellites. The feasibility and methods of US satellite programs are also mentioned.

88 The page discusses the concept of space stations and their feasibility. It mentions that space stations are not considered feasible due to their high cost, despite not violating any physical laws. The conversation also touches on the Russian space program and their literature.

89 The content of the page appears to be a mix of random words and phrases that do not form a coherent summary.

90 The page discusses the development of the hydrogen bomb and the Russian space program. The speaker mentions a lack of cooperation from the Russian government and the importance of access to statistics and information for making judgments on space programs.

91 The page discusses the challenges faced in obtaining accurate information on Soviet space technology due to lack of cooperation from Russian sources. It also touches on the broader issue of information security and the need-to-know principle.

92 The page discusses the tracking of Russian books and the importance of cooperation in gathering information. It also mentions the presence of witnesses from the Air Force.

93 The page contains statements from various individuals involved in intelligence and research related to space tracking. They discuss the exchange of information and provide short presentations on Soviet space programs.

94 This page discusses the tracking and observation of space vehicles by the Air Force. It mentions specific locations and organizations involved in the research. The conclusion is that the observations confirm the presence of debris in space.

95 The page discusses the lack of cooperation and restraint in releasing data related to evaluations. It also mentions the possibility of restrictions on certain individuals. The conversation touches on the accuracy of Soviet statements about their space program.

96 The conversation is about tracking a mission to the moon and the equipment needed for it. There is uncertainty about the exact distance reached and what occurred after reaching a certain point. There is also discussion about the equipment and capabilities of Russian scientists.

97 The page discusses Soviet missile capabilities, specifically their range and accuracy. It also mentions the timing for launching missiles and questions the accuracy of Soviet claims.

98 The page discusses a closed session hearing about the Russian position on missile and space advancements, including their superior boosters and engines. It also mentions technical advances in dams and reservoirs.

99 The page discusses the comparison between the US and the Soviet Union in terms of guidance technology and miniaturization in the missile and rocket industry. It suggests that while the US may be behind in some aspects, they are not hopelessly behind. The need for a closed session to provide more information on Russian missile guidance is mentioned.

100 The page discusses guidance systems for missiles and the tracking of a missile's trajectory. It mentions the superiority of Russia in ground station control and tracking networks. The conversation also touches on the velocity and position of the missile when tracking ceased.

101 The conversation discusses the lack of information regarding the trajectory of a rocket that was being tracked. They speculate about its potential location and whether it entered the moon's gravitational field. It is uncertain if the rocket returned to Earth or continued into space due to limited tracking data.

102 The committee discusses the possibility of an unknown impact and the need for further investigation, leading to a decision to go into executive session.

103 Dr. Hans Ziegler testifies before a committee about Soviet space technology and his involvement in research during a specific time period.

104 The page discusses the author's activities and involvement in the United States Signal Corps, particularly in relation to tracking and communication equipment for satellite launches. It mentions the use of various types of antennas and the importance of high sensitivity and low noise equipment.

105 The page discusses the availability and operation of tracking stations for special projects and space probes. It mentions the continuous recording of satellite orbits and the ability to reproduce observations. It also highlights the challenges of changing frequencies and the need for different adapters. The page concludes by mentioning ongoing activities and collaborations with other tracking organizations.

106 This page discusses the availability of prepared feeds and the ability to identify signals on certain frequencies. It also mentions the difficulties in identifying signals from ground stations and the challenges of frequency shifts.

107 The page contains a jumbled collection of letters and symbols that do not form coherent sentences or meaning.

108 The page discusses the wavelength and frequency requirements for a specific type of antenna being used by the Russians. It mentions that the facility is adaptable to different frequencies and that the reception sensitivity will be similar to that of a previously used dish. The page also mentions the inability to determine the direction and frequency of incoming signals.

109 The page discusses the detection of signals and changes in frequencies, but is inconclusive in determining their origin or significance.

110 The page discusses information related to Russian frequencies and identifications during a specific time period. It mentions the lack of certain identifications and suggests that further discussion is needed. The speaker emphasizes that they do not have definitive conclusions but rely on available information.

111 The page discusses the inability to draw a positive conclusion from certain information and the importance of considering other factors. It also mentions the use of artificial Doppler and the need for precise conclusions.

112 The page discusses a conversation about observations and simulations related to a certain phenomenon. The participants discuss the lack of evidence and the difficulty in comparing different observations.

113 The page contains a mixture of words and phrases, but without context or structure it is difficult to determine the exact meaning or topic.

114 The page discusses conversations about recent developments in space technology, including the involvement of important figures and the use of certain equipment by the Russians. It also mentions the possibility of them making moonshots.

115 The page discusses the possibility of going beyond the moon and the accuracy of guidance systems. It also mentions the launch of the Mechta by the Soviets and the simulation of certain scenarios. The conclusion is based on the available knowledge from the US government.

116 The page discusses the transmission of signals from a vehicle that traveled to the moon and then disappeared. The accuracy of the Soviet Union's data is uncertain, but it is believed that the vehicle would have entered a solar orbit. The possibility of simulating the lunar signals is discussed, as well as potential reasons for the failure to detect signals with a radio telescope.

117 The page contains a conversation between individuals discussing their interactions with scientists in Moscow and the information they were able to gather.

118 The page contains a conversation about someone needing to speak with different individuals and their opinions on the matter. It also mentions a statement that someone wants to make during their testimony.

119 The page discusses a conversation between Mr. Miro and someone else about the Russian government's claim to be the first people to orbit the Earth. There are also mentions of a vehicle orbiting the sun and Mr. Fulton's statement.

120 The page discusses the trajectory of a vehicle and the correlation of figures based on data.

121 The page discusses the possibility of the Russians having a "Hine of the sun" and suggests that if this is true, then the Americans should also be credited for reaching the sun. It mentions Operation Turi and Operation Farstds as potential sources of more information. It concludes by stating that definitive proof cannot be provided based on tracking alone and that the quality of the tapes is inadequate.

122 The page discusses the need for proof regarding a certain operation and whether it went into orbit. It also mentions the Defense Department's stance on Soviet space technology and the desire to improve US capabilities. The conclusion asks for a comparison between two operations.

123 The page contains data about Operation Farside, but it is difficult to decipher due to the formatting and lack of context.

124 The content of the page has been redacted and is not available for review.

125 The subcommittee is meeting to determine the accuracy of classified information regarding Soviet space technology and to uncover any additional secrets. Testimonies will focus on the potential threat to global peace posed by the Soviet Union.

126 The page discusses a story titled "The Big Red Lies" that appeared in True magazine. It questions the existence of the Soviet moon rocket and provides evidence to support the claim that it is propaganda. The page also mentions President Eisenhower's commendation of the Russian achievement and the successful launch of Project Score.

127 The page discusses the need for open dialogue and understanding with the Soviet Union, as well as the importance of recognizing their scientific achievements. It also mentions an author who has evidence to debunk misconceptions about Russia's technological advancements.

128 The page discusses statements made by various individuals regarding space technology and satellite research. It mentions previous testimony and errors of judgment made by Dr. von Braun.

129 This page discusses the evidence and proof related to a photograph taken in England, which is being used as evidence in a case. It also mentions radio tracking stations and their inability to detect signals from a certain source.

130 The page discusses the need for consensus on whether someone was present or not and references Goldstone tracking reports. It also mentions mysterious signals and raises doubts about the existence of lunik.

131 The page discusses various aspects of an unidentified subject, including obstacles, accomplishments, physical characteristics, and potential abilities.

132 Signals allegedly received from Lunik all, recorded by various individuals in Moscow and Stalingrad.

133 This page contains information about the transportation of goods from Datuk to the Castes part of the Bovis Union. It also mentions the involvement of various individuals and organizations in this process.

134 The page discusses the weight of Sputnik III and whether the Soviets had enough thrust to send a rocket to the moon. It is argued that thrust alone is not enough for a successful mission, and proof is lacking regarding the weight of Sputnik III. The importance of a proper guidance system is emphasized, and it is suggested that the Soviets did not have the necessary technology for accurate navigation.

135 The page discusses the Soviet Academy of Sciences and their involvement in the Soviet space program. It mentions the development of an inertial guidance system and doubts about its accuracy. There is also a mention of controversial announcements regarding the use of IK by certain individuals.

136 The page discusses the alleged photo references of an amateur photographer and the temperature requirements for certain types of film. It also mentions an interview with a physics department chairman about Soviet space technology and the ability to launch a rocket to the moon.

137 The page discusses the sporadic signals heard by Goldstone radio telescope and explores possible sources for those signals, including the movement of a moon. It also mentions other stations and repairs being done.

138 The page discusses alleged signals intercepted by a plane flying over the ocean and speculates on their origin, including the possibility of signals bouncing off the moon or other airplanes. It also mentions bursts of signals emitted from the 10 to 3 megahertz range.

139 This page contains a released CIA document from 2004.

140 The page discusses the recognition and interpretation of signals from Jupiter, including the possibility of advanced signals being diffracted onto Earth. It also mentions the case of Goldstone and the challenges they faced in operating at different frequencies. The issue of whether certain signals should be made public is raised, considering the potential for classified information.

141 The page discusses the possibility of a launched rocket traveling 3,000 miles to Goldstone station in Hawaii. It also mentions the need for further data and questioning surrounding the topic.

142 The page discusses the accuracy and precision of equipment used by scientists for space exploration, specifically regarding the perigee altitudes of three space probes. The author expresses skepticism about the claimed 9-mile accuracy and suggests that the scientists may not have had accurate measurements.

143 The conversation is about the weights of rockets launched by the Russians and the suspicion that they may not be accurate. The scientists discuss the possibility of using optical equipment to determine the true weight of objects in space. There is also mention of tracking issues with satellites and the configuration of the rockets.

144 The page discusses a Soviet delegate's statement about the weight of the Sputnik satellite, which was initially reported incorrectly and later corrected. It also references an article about the distribution of electron counts and positive ions in the atmosphere. The development of meteorological rockets is mentioned as well.

145 This page features a conversation between Mr. Magar and Mr. Kon regarding data comparison between American and Russian papers. Mr. Kon defends the accuracy of American data while Mr. Magar argues that Russian papers are often not classified and readily available to Russians. The conversation is interrupted, and another witness from Colorado is mentioned.

146 The page discusses comparing results and data between different sources, specifically in relation to meteorological rockets and research into the upper atmosphere. It highlights the importance of accurate and reliable data in this field.

147 The page discusses equipment and information, but the content is unclear and disjointed. There are references to accumulated equipment, impressions, and the background of certain individuals. It is suggested that there may be some sort of investigation or inquiry taking place.

148 The page contains a conversation between individuals discussing the credibility of sources and the expertise of scientists in relation to a particular topic. Quotes from experts are provided to support different viewpoints.

149 The page discusses the tracking of military signals and the classification of documents. It also includes a conversation about travel to Europe and the surrender of a passport.

150 The page discusses a conversation about someone surrendering their passport and their departure from the United States to Spain. It also

mentions an organization called Norte Control which is involved in radio communication.

151 The page contains a conversation between individuals discussing their experiences in war and their backgrounds. It also mentions the International Brigade and the Abraham Lincoln Brigade.

152 The page is a jumble of words and letters that do not form a coherent summary.

153 The page contains a mix of random letters and words with no clear meaning or purpose.

154 The page contains a conversation transcript discussing the possibility of a hidden message in a computer program.

155 The page discusses attempts to contact Dr. Pickering, the signals being tracked, and the difficulty in deciphering them. It also mentions an anecdote about a stolen document and the challenges of finding hidden information.

156 The page contains a conversation between Mr. Larkin and Mr. Marsa about a lunar affair. They discuss the significance of a phone bill and the timeline of an article.

157 The page is a transcript of a conversation between individuals discussing the credibility of information provided by government agencies and the possibility of scientists lying. The speaker disagrees with the notion that scientists would lie about their work but suggests that it is conceivable for government officials to do so.

158 The page contains a conversation about American scientists and their credibility, as well as discussions about equipment and technology. There are mentions of lying, opinions, and the need for proof.

159 The page contains a conversation about American scientists and their alleged ties to the Russian government. The speaker questions the credibility of these claims and argues that not all scientists are influenced by political affiliations.

160 The page contains a conversation between individuals discussing the potential impact of a Russian individual on the space program and the possibility of propaganda efforts. It is suggested that such efforts would be easily recognized and unlikely to succeed.

161 The page discusses a conversation between Mr. Kuan, Mr. Mallan, and several scientists regarding the accuracy of atomic clocks and their comparison to quartz clocks. The scientists explain that atomic clocks are more precise, measuring accuracy within a millionth of a second.

162 The page discusses a conversation about the importance of precise timing and technology in tracking Russian space activities, particularly their use of atomic clocks and mechanical printers for accurate measurements. It suggests that the Russians may have advanced technology beyond what is known.

163 The page discusses the responsibility and decision-making process regarding military and scientific preparedness. It emphasizes the need for the United States to maintain leadership in scientific and technological advancements, particularly during times of war.

164 The page discusses the Cold War and the impact of information on dictators. It emphasizes the importance of accurate information and

criticizes the spread of misinformation. A witness named Dr. Alan I. Shapley from the Bureau of Standards is mentioned.

165 The page is a transcript of a testimony given by Alax K. Sapley, Assistant Chief of the Radio Propagation Prison Division at the National Bureau of Standards. He discusses their role in recording solar emissions and explains that they are not a tracking station but a scientific research station.

166 The page discusses a series of questions related to the timing of an event and requests for computations to be submitted for the record. The purpose of these questions is to determine the time when certain signals were received.

167 The page discusses the calculations and positions of the Moon and Jupiter. It suggests that what was thought to be lunar signals may have actually come from Jupiter. The separation between the two celestial bodies is given for different dates. The angle between them on January 5 is mentioned. The location of the tracing station and the possibility of signals coming from Jupiter are also mentioned.

168 The page discusses various factors involved in measuring the frequency and signals from Jupiter. It concludes that the signals are unlikely to come from any other natural source and suggests that scientists prefer to speak in terms of probabilities rather than absolutes. There is also mention of measuring the Moon's radioactivity and magnetic field.

169 The page discusses errors in the Jodrell Tankwl telescope installation and the need for specific information to determine the accuracy of tracking and crossbearings. It also raises questions about the necessary level of accuracy and the ability to derive location, speed, and distance from the data.

170 The page discusses the possibility of deriving location, speed, and distance using radio signals and Doppler effects. It also mentions the need for assumptions and simulations in order to obtain accurate information.

171 The page contains a jumbled mix of text that is difficult to decipher and understand. It appears to be related to simulations, tracking satellites or space vehicles, and the Jodrell Bank observatory. The content is unclear and lacks coherence.

172 The page discusses the terms and conditions of a document related to Rasta clothing, emphasizing the importance of proper signature and following regulations. It also mentions the tremendous power of communication and the role of the government in regulating certain activities.

173 This page contains a document with various codes, abbreviations, and technical terms related to computer systems and operations.

174 The page discusses the importance of accuracy and attention to detail in a project, emphasizing that even the slightest mistake can lead to failure. It also mentions the need for guidance components in an engine and the recent computation of their impact.

175 This page discusses the Soviet optical racing network and its use of servos as the computer's hands.

176 The page discusses the radio emissions from Jupiter and the possibility of receiving these emissions on Earth. It mentions the frequencies at which

these emissions occur and the power of the telescopes used to detect them. It also mentions the bending of radio waves by ionized clouds in nearby space.

177 The page discusses the importance of bonding by ionized clouds in the atmosphere and speculates on its potential effects. It also mentions the observations made by Russians and concludes that further study is needed.

178 This page contains a discussion about signals from Jupiter and the possibility of detecting harmonies in those signals. Different scientists have different opinions on the matter, with some suggesting that the signals can be detected and others being skeptical. The discussion also touches on the idea of refracted or bent signals and the variations in data depending on external factors. Ultimately, it is concluded that there are no definitive deviations in the signals at this time.

179 The page contains a conversation between Mr. Marga and Dr. Shapley about the limitations of scientific observation and the impact it has had on the progress of physics. They discuss the possibility of certain conditions contradicting accepted theories and conclude the meeting.

180 The page is a transcript of a subcommittee meeting discussing Soviet space technology. Lloyd Mallax, author of "The Big Red Lie," is questioned about his knowledge of certain towns in Poland. He is unsure and recalls that his father left before the USSR took over that part of Poland.

181 Mr. Mallan appeared voluntarily before the committee and has been cooperative. There was a discussion about the spelling of his name on his passport, but he insists it was spelled correctly. He does not understand Russian but can recognize some words. None of the individuals he contacted before leaving the country contacted him upon his return, and one of them has died.

182 The page is a transcript of an interview discussing someone being killed in Spain and the narrator's travels through Germany. The narrator was interrogated for several months and volunteered information to the authorities.

183 During a conversation, Mr. Maras discusses the coverage of a company and the places he went to gather information. Mr. Mallan asks if any information was withheld, and Mr. Maras confirms that there was. Mr. Slavin then introduces himself as the director of Project Space Track.

184 This page is a statement given by Mr. Rover M. Slavin, Director of Protect Space Track, about the Russian Tuna probe and their work in tracking objects in space. The data is collected from a network of observers and processed at the Space Track Control Center. The statement also mentions the success of Goldstone and Stanford groups in obtaining tracking data.

185 The page discusses the possibility of simulating lunar signals on a handheld device. It mentions that the signals heard were strong, except for one exception, and that some observers reported signal sharing. The page also explores the idea of airborne transmissions and the potential involvement of satellites.

186 The page discusses the possibility of simulating aircraft and ground movements using signals. It mentions the use of antennas and the

potential to measure distance and speed based on frequency changes. There is also a mention of data observations from Goldstone and Russian sources.

187 The page discusses the analysis of astronomical measurements, specifically the position of the moon and a probe. It mentions the agreement between the Russian announcements and the measured positions. The assumptions involved in estimating speed and distance are also discussed.

188 The page contains a jumbled mix of letters and words that are difficult to understand or make sense of.

189 The page discusses the possibility of Russian submarines waiting in international waters to transmit radio signals to the United States when the moon is above the horizon. The accuracy of this claim is questioned, and it is mentioned that recordings of 30-megacycle signals were obtained at a later time.

190 The discussion on the page revolves around the possibility of a hoax and the effort required to simulate a signal across the sky. The witness has been involved in satellite tracking for over 20 years. There is a mention of accuracy in Soviet announcements about lunar missions.

191 The page discusses the importance of launch data in accurately determining the distance and location of space probes. It emphasizes that launch data is crucial for tracking systems to compute and correct for errors in determining a probe's distance and trajectory.

192 The page discusses the differences in measurements between being on Earth and being in space, specifically in relation to satellite altitudes. It also mentions the importance of knowing the position of the moon in certain scenarios.

193 The page discusses a conversation about the trajectory of a Russian probe to the moon and the possibility of its signals coming from the moon itself. The analysts rule out the signals being a reflection and discuss the different possibilities for the probe's trajectory.

194 The page discusses various trajectories and parameters related to lunar exploration. It mentions the background of individuals involved in the business and raises questions about signal reception at a certain distance from Earth.

195 The page discusses radio signals and refraction, specifically in relation to the Soviet Union and the United States. The author mentions the reception of signals and their dependence on certain conditions. The page also references research and development, as well as air traffic.

196 The page contains a conversation between multiple individuals discussing topics related to intelligence, security, and monitoring.

197 The page discusses a conversation about Jodrell Bank and guidance for a moon shot, addressing issues such as velocity error and the size of the spacecraft.

SOVIET SPACE TECHNOLOGY

HEARINGS

BEFORE THE

COMMITTEE ON SCIENCE AND ASTRONAUTICS AND SPECIAL SUBCOMMITTEE ON LUNIK PROBE U.S. HOUSE OF REPRESENTATIVES

EIGHTY-SIXTH CONGRESS

FIRST SESSION

MAY 11, 12, 13, 14, 28, 29, AND EXECUTIVE SESSIONS OF MAY 12 AND 14, 1959

[No. 46]

Printed for the use of the Committee on Science and Astronautics

UNITED STATES
GOVERNMENT PRINTING OFFICE
48438 WASHINGTON : 1959

Approved For Release 2004/08/31 : CIA-RDP63T00245R000100290011-7

COMMITTEE ON SCIENCE AND ASTRONAUTICS

OVERTON BROOKS, Louisiana, *Chairman*

JOHN W. McCORMACK, Massachusetts
GEORGE P. MILLER, California
OLIN E. TEAGUE, Texas
VICTOR L. ANFUSO, New York
B. F. SISK, California
ERWIN MITCHELL, Georgia
JAMES M. QUIGLEY, Pennsylvania
DAVID M. HALL, North Carolina
LEONARD G. WOLF, Iowa
JOSEPH E. KARTH, Minnesota
KEN HECHLER, West Virginia
EMILIO Q. DADDARIO, Connecticut
WALTER H. MOELLER, Ohio
DAVID S. KING, Utah
J. EDWARD ROUSH, Indiana
JOSEPH W. MARTIN, Jr., Massachusetts
JAMES G. FULTON, Pennsylvania
GORDON L. McDONOUGH, California
J. EDGAR CHENOWETH, Colorado
FRANK C. OSMERS, Jr., New Jersey
WILLIAM K. VAN PELT, Wisconsin
A. D. BAUMHART, Jr., Ohio
PERKINS BASS, New Hampshire
R. WALTER RIEHLMAN, New York

CHARLES F. DUCANDER, *Executive Director and Chief Counsel*
Dr. CHARLES S. SHELDON II, *Technical Director*
SPENCER M. BERESFORD, *Special Counsel*
PHILIP B. YEAGER, *Special Consultant*
HARNEY S. BOGAN, *Staff Consultant*
JOHN A. CARSTARPHEN, Jr., *Counsel*
RICHARD P. HINES, *Staff Consultant*
Lt. Col. PAUL B. SCHUPPENER, *Staff Consultant*
RAYMOND WILCOVE, *Staff Consultant*

SPECIAL SUBCOMMITTEE ON LUNIK PROBE

VICTOR L. ANFUSO, New York, *Chairman*

LEONARD G. WOLF, Iowa
JOSEPH E. KARTH, Minnesota
DAVID S. KING, Utah
FRANK C. OSMERS, Jr., New Jersey
WILLIAM K. VAN PELT, Wisconsin

SPENCER M. BERESFORD, *Special Counsel*

CONTENTS

SOVIET SPACE TECHNOLOGY

MONDAY, MAY 11, 1959

House of Representatives,
Committee on Science and Astronautics,
Washington, D.C.

The committee met at 10 a.m., in the caucus room, Old House Office Building, Washington, D.C., Hon. Overton Brooks (chairman) presiding.

The Chairman. The committee will please come to order.

This morning the committee opens a new hearing. We will meet here as long as the attendance at the hearings justifies meeting in the caucus room, provided we get permission to meet here.

The purpose of the hearing now is to establish the authenticity of Soviet claims as to the sputniks and lunik and also to dig into the status of scientific advancement in other countries of the world, especially in the Soviet Union.

We have a number of witnesses scheduled. The hearings are scheduled to last a week. If we can move the witnesses up, however, and hear them faster than we now expect to be able to do, we can finish the hearings earlier than that. It will depend on the number of questions and the number of facts that we are able to develop in the course of the hearings.

I want to first ask the members of the committee if they can hear me. To start with we have Mr. Lloyd Mallan, who is the author of the story entitled "The Big Red Lie" which was published in True magazine.

I want to recognize Mr. Anfuso to make a short statement at this time.

Mr. Anfuso. Thank you, Mr. Chairman.

Mr. Chairman, I should like to make it clear at the outset of these hearings that the representatives of True magazine which published this story "The Big Red Lie," Mr. Ralph Daigh, and its author, Lloyd Mallan, a very reputable reporter, both asked for this investigation, and you were good enough to grant it in order to set the record straight once and for all.

I believe, in the best interests of the American people, this story should be fully investigated, and I am glad that you acquiesced in that. I can assure you, Mr. Daigh and Mr. Mallan, that while I cannot be here for all of the hearings, this committee will be extremely fair in the interrogation of the witnesses and you will be given every opportunity to state your case. I am confident that you are motivated by the best interests of the country in printing and circulating this article.

I am equally confident that the same spirit of patriotism will lead you to retract your story and to give as much publicity to the retraction, should it develop in these hearings that the story was unwarranted. Am I correct in stating that, Mr. Daigh and Mr. Mallan?

Mr. MALLAN. Yes, you are.

Mr. ANFUSO. Thank you, Mr. Chairman.

The CHAIRMAN. Thank you, Mr. Anfuso.

Before proceeding with the testimony of Mr. Mallan, I wish to say that we have two important telegrams that the committee has received and I am going to ask Mr. Beresford, special counsel for the committee, to read those two telegrams.

Mr. BERESFORD. The first is a telegram from the Stanford Research Institute Station at Menlo Park, Calif. The telegram is dated May 5.

> Stanford Research Institute was notified on January 2, 1959, of the launch of a Russian moon probe. A telephone check with Project Space Track, Air Force Cambridge Research Center revealed that the official frequencies used were 19.993, 19.995, and 19.997 megacycles. Furthermore, it was suspected that a frequency of 71.2 megacycles might be used. Preparations were made at the Stanford Research Institute field site to monitor these four frequencies. This monitoring began at approximately 0030 universal time on January 3. The receiving equipment consisted of the following: Antennas, two 2-element yagies were used at 20 mc. One antenna was vertically polarized, the other horizontally polarized. Each antenna uses a gamma match to 50 ohm RG 8U, equal length of cable feed a hybrid loop. Add and subtract output of loop, are connected through equal lengths of cables to receivers. The yagies were movable, 360 degrees in azimuth. The 71.2 mc. antenna is a size element yagi connected directly to a NEMS-Clark receiver. The 20-megacycle receivers were two Collins 51 J-4's tuned to approximately 19.993 megacycles. B.F.O. was normally on, AVC off, Limiter off, BFO filter on 1 kc., crystal filter on positions 2 to 4. Signal was recorded on an ampex F.R. 1102 converted to F.R. 1104, four-channel record reproduce machine on half inch tape at 15 IPSW was used. WWV was recorded on first channel. The 20 mc. signals from the vertically and horizontally polarized channel were recorded on channels 2 and 3. The 71.2 megacycle signal was recorded on channel 4. The following are notes and comments:
>
> The signal was expected around moonrise January 3 which occurred approximately at 0915 universal time. Signal was not picked up until 1045 universal time and was recorded on magnetic type until 1128 u.t. Monitoring continued until approximately 1300 u.t. The signal was very weak at the beginning and at no time was it stronger than weak. It appeared to be CW and appeared to follow the moon by about an hour and a half. There were two types of fading on the signal. One was generally deep fades in both channels, the other was polarization fading. The signal shift from vertical to horizontal polarization and back, etc. It was noted after about 2 hours that the apparent Doppler shift had reached a value of 1.5 kc. Comments by trained observers were made to the effect that this shift was too light for satellite on the way to the moon. It was attempted to obtain contact with the moon probe on January 4 without success. Footnote the two Collins receivers had been turned on a minimum of 15 hours. The signal required almost continuous tuning for peaking. It was too weak to obtain any factual Doppler shift. Antennas were rotated frequently to maximize signals. One-minute polarization fades. Longer periods for deep fades. Nineteen hundred and fifty-eight Delta-2 recorded intermittently on 20.004 mc. between 1110 and 1120 u.t.

The other telegram is from Dr. Lovell, the director of the Jodrell Bank Radio Telescope Laboratory in England dated May 8, 1959:

> We did not succeed in detecting any signals from lunik. Search was made with the large radio telescope over a frequency band covering 183.6 megacycles which was the stated tracking frequency during the night of January 3 to 4 when lunik was said to be close to moon. No tracking errors could account for failure because telescope beam was adequate to cover position of probe at time of close approach. Telescope was in similar operational condition which has enabled us to track Pioneer to 400,000 miles east. We conclude that lunik was not transmitting continuously at least on this frequency during that night.

Emphasize that we do not disbelieve in existence of lunik. That frequency might have been inoperative or some special ground command or other features that are in existence in America might make it impossible without special apparatus. We were not forewarned in spite of previous close association with Russian scientists responsible, neither have we received any satisfactory explanation from our inquiries by letter to Russia or from the Soviet delegate at The Hague meeting of Cospar.

The CHAIRMAN. Have you finished them both?

Mr. BERESFORD. Yes.

The CHAIRMAN. Now, Mr. Lloyd Mallan is a writer for True magazine. I do not have the date this article came out. What date was that, Mr. Mallan?

Mr. MALLAN. The date, Mr. Chairman, was April 21.

The CHAIRMAN. On April 21 True magazine carried an article entitled "The Big Red Lie," by Lloyd Mallan. The purpose of the article casts grave doubt on whether this was a big Red hoax or whether it had any elements of the truth in it so we have asked Mr. Mallan to come here this morning and give us his idea of the whole situation supporting the view that he had taken on the matter. He will be followed by the other witnesses that we have scheduled.

I think, because of the nature of the inquiry, the witnesses should be sworn.

Mr. FULTON. Will the chairman yield on a jurisdictional point? The question comes up on these hearings as to whether they will be limited to the point of proof or the disproof of the particular missile flight or whether they will be extended under the questioning on the hoax to the general field of Russian progress in the missile field. My own feeling is if we are going to proceed on the question of developing the truth or falsity of the hoax that we should do so and not under that particular heading spread it to a general investigation of the Russian progress in this field. I would, therefore, respectfully request that the hearings be aimed at the hoax and that if we, as some of the newspaper accounts have said on Sunday, quoting the chairman, are to proceed with an investigation of the Russian progress, that we do that independently of this particular so-called magazine story on the hoax.

The CHAIRMAN. The Chair as a rule allows the members of the committee broad latitude in their questions. I would not like to be put in the position of cutting Mr. Fulton or any other member of the committee off if they stray a little bit from the story. It is the pleasure of the committee. We have some witnesses, however, that want to testify on the broad field of Russian progress in science. We have General Boushey to whom we have allotted a full day who will be here. What is his name?

Mr. FULTON. Might I ask a question of my colleague Mr. Anfuso, who heads the Subcommittee on International Cooperation. Is it your intention to have any hearings on the broad field that the chairman and I have outlined?

The CHAIRMAN. I have a schedule mapped out which I would like to take up in executive session on the subcommittee program.

Mr. FULTON. Could I preliminarily, then, make a motion that we for 2 days schedule hearings as a full committee on the question of this hoax and then have for later decision how we will head the Subcommittee on International Cooperation.

The CHAIRMAN. Will you read the list of witnesses we have, Mr. Ducander?

Mr. Beresford, do you have a list of all our witnesses?

Mr. MILLER. Mr. Chairman, I would like to second Mr. Fulton's motion.

Mr. TEAGUE. We can't tell where this is going to lead, Mr. Chairman. I do not concur in this motion. I think we should see just where it leads.

Mr. MILLER. My only thought in seconding the motion is if we are going to go into the relative field of the Russians in this field, I would like to see the witnesses. I would like to go into executive session and get some of our own people before us who I think are quite competent in making this judgment and not bring in a lot of people who will come here and volunteer statements that might be in contradiction to what our own scientific staff within the Government may have and make this a sounding board for a lot of propaganda.

Mr. FULTON. Would the gentleman from Texas yield?

The CHAIRMAN. Will you get a list of the witnesses? We had a list of the witnesses read into the record last week but I think we ought to have a list of them now.

Mr. FULTON. Let us separate the jurisdiction of the subcommittee. That is my point. Where the subcommittee has jurisdiction of international cooperation I feel that that portion of the investigation should be under the subcommittee. If the whole committee wants to take it, then it should be done separately and not in passing on the validity of a hoax of a magazine article.

I think the two should be separated and I am respectfully requesting that that be considered at this time so that we have the general course of what we are going to do in these hearings laid out, that the committee knows and my resolution is a suggestion.

Mr. TEAGUE. Well, the chairman stated that he planned to have an executive session to set out these subcommittee problems and surely we should not do it this morning.

It should be on an overall basis.

Mr. FULTON. Then why should we not limit our hearings at this point to simply the hoax and not to the general field of Russian progress for the time being?

Mr. MILLER. If the gentleman will yield, I will withdraw my second to his motion, because I think that the committee will spend most of the day hearing Mr. Mallan and other proponents, and then, as I understand it, this executive committee meeting will be held. When?

The CHAIRMAN. Whenever we can.

Mr. MILLER. I think, if the executive committee meeting is going to be held within the near future, before we get further into this thing, that will take care of it. Otherwise I shall reserve the right to make that resolution at some future time. I would like us to get our own housekeeping straightened out before we get far afield.

Mr. ANFUSO. Mr. Chairman.

The CHAIRMAN. Mr. Anfuso.

Mr. ANFUSO. The only time I could attend an executive hearing would be today, perhaps sometime this afternoon; then I will be gone the balance of the week.

The CHAIRMAN. If we finish with Mr. Mallan early enough, we can hold it right after we finish with him, just before 12 o'clock.

I do not think we should take much time on that.

Mr. FULTON. That is satisfactory to me.

Mr. MILLER. Will you withdraw your motion?

The CHAIRMAN. The motion is withdrawn. The second is withdrawn.

We will proceed, then.

Mr. Mallan, will you take the oath?

Do you solemnly swear that the testimony that you shall give this committee in matters now under consideration will be the truth, the whole truth, and nothing but the truth, so help you God?

Mr. MALLAN. Yes; I do.

The CHAIRMAN. Do you have a prepared statement, Mr. Mallan?

Mr. MALLAN. No, sir; I do not.

The CHAIRMAN. Well, would you like to proceed in your own way?

STATEMENT OF LLOYD MALLAN, NEWS REPORTER, TRUE MAGAZINE

Mr. MALLAN. I would like to start by recalling the cable from Dr. Lovell. I thought it was exceptionally significant for the simple reason that Dr. Lovell, who is the foremost radioastronomer and an outstanding expert on radiotelescopes, mentioned that his instrument, the largest mobile radiotelescope in the world or in the free world, was perfectly tuned and on the 3d and 4th of January tried to find signals on 183.6 megacycles from the lunik. He failed to find these signals and his explanation, as I recall the cable, was that undoubtedly the lunik transmitter on that frequency was not operating.

Yet on January 4 at the Goldstone Dry Lake tracking station on 183.6 megacycles the jet-propulsion laboratory found signals on that frequency with an 85-foot diameter radiotelescope, which is very efficient, by the way.

Mr. ANFUSO. Excuse me, Mr. Chairman.

The CHAIRMAN. Mr. Anfuso.

Mr. ANFUSO. May I say this: Don't you think, Mr. Mallan, you are speaking now like an expert? Don't you think that in the best interests of your case that you ought to first give us something of your background?

Mr. MALLAN. Yes, Congressman.

The CHAIRMAN. And why not also add to it the facts leading up to the writing of the article there.

Mr. ANFUSO. That is right.

The CHAIRMAN. So that we will have it in chronological order.

Mr. MALLAN. I will try to make this as brief as possible. My background in astrophysics, which includes radioastronomy and nuclear physics, astronomy in general, missile research and development, aviation research and development, space physiology, and a great many other subjects on a high level, was acquired on a self-trained basis, on the basis of observations at our own research centers and on the basis of 20 years' experience as a newspaperman as well as in the beginning a layman intensely interested in all these aspects of science. I went to the Carnegie Institute of Technology but in order to make this briefer, in the fall of 1957, after the Soviet Union launched their first sputnik, Fawcett Publications and True magazine offered to send me to the Soviet Union to find out or to acquire a behind-the-scenes story of the Soviet Union.

The free world spontaneously, almost as a whole, took the attitude that the Soviet Union was 'way ahead of the United States and I had that attitude before entering the Soviet Union.

When I got there, day after day I received shock after shock because I went there knowing what we were doing, knowing the type of research we had engaged in for the past decade or longer in the field of space medicine, knowing our research in high altitude and high-speed flight with rocket-powered aircraft.

In fact, I have published several books on the U.S. air research and development effort, also the air defense effort. So, I expected the Soviets to have the equivalent, at least, if not something far better, than we had. I forgot to mention that I carried with me letters of introduction to Soviet scientists, to the leading Soviet scientists, many of the leading Soviet scientists, from their counterparts in the United States.

My first interview was with the head of the Soviet Space Flights Commission, and I was startled when I asked him about radiotelescopes which are very necessary for radiotracking, especially over great distances such as through outer space———

Mr. TEAGUE. Mr. Chairman.

The CHAIRMAN. Mr. Teague.

Mr. TEAGUE. Mr. Mallan, do you speak Russian?

Mr. MALLAN. No; I do not.

Mr. TEAGUE. How did you interview them? How did you interpret?

Mr. MALLAN. I had a girl interpreter who knew no science.

Mr. TEAGUE. The interpreter or yourself?

Mr. MALLAN. Well, she was a member of the Soviet Intourist Bureau. However, several of the scientists spoke broken English and the jargon of science is international and many times I corrected her when she was interpreting things in too literal a manner and not understanding the words herself in Russian. I also made tape recordings of a number of these interviews with Russian scientists.

Mr. FULTON. Could we have those submitted for the record, Mr. Chairman?

The CHAIRMAN. Do you object to submitting those for the record?

Mr. MALLAN. I do not.

The CHAIRMAN. Do you have them there?

Mr. MALLAN. I do. These are only a portion of them, however. I have transcripts of about half a dozen of them.

Mr. FULTON. I move that the transcripts and tape recordings be put in the record.

The CHAIRMAN. It does not require a motion. If there is no objection, they will be placed in the record.

Mr. MILLER. At this point I also ask that Mr. Mallan place in the record the names of the scientists in this country who gave him the letters of introduction to the scientists in Russia.

Mr. MALLAN. I would be most happy to do that, sir.

(The legible transcripts and list of scientists referred to follow, the remainder are in the committee file:)

INTERVIEW IN MOSCOW WITH PROF. YURY POBEDONOSTEV AND PROF. KIRIL PETROVICH STANYUKOVICH, ARRANGED BY ANATOLY KARPENKO, PERMANENT SECRETARY OF THE PERMANENT COMMISSION ON INTERPLANETARY COMMUNICATIONS

(Interview by Lloyd Mallan, American author)

Comment: This is an interview at the Academy of Sciences, U.S.S.R., arranged by the scientific secretary, Karpenko. Karpenko is scientific secretary of the Permanent Commission on Interplanetary Communications of the Academy. This meeting took place on the day before May Day, April 30, 1958. Present at the meeting were: Anatoly Karpenko, Prof. Yury Pobedonostev, and Prof. Kiril Petrovich Stanyukovich. Pobedonostev is professor of aeronautical sciences at the Moscow Aviation Institute. He also claimed to be the pioneer in flying aircraft with liquid-rocket assist back in the early 1930's. Professor Stanyukovich is known as a specialist in gas dynamics and is a professor at the Moscow Technological Institute. He is also the head of the mathematics department of that institute. Karpenko had arranged this interview but surprised us with both Pobedonostev and Stanyukovich; we had expected to be interviewing Federov, who Karpenko had told us was head of the propulsion institute, and who recently turned up in Geneva as one of the Soviet scientists consulting on the control of the setup of the control system for the prevention or detection of nuclear form experiments. At any rate, we were pleasantly surprised because both Stanyukovich and Pobedonostev are allegedly top experts in their fields. As we entered the Academy, Pobedonostev was there waiting. Stanyukovich had not yet shown up. Pobedonostev greeted us warmly and shook hands. When I asked whether we might tape record the interview with him (through our interpreter) he consulted with Karpenko and then said something to the interpreter which she translated to us as "No, he would not like to be taped during the interview." He felt that using a pencil was much better. Although this was the day before May Day, one of the biggest holidays in the Soviet Union, these men were apparently sincerely interested in helping us gather information for our book. Karpenko had told us over the telephone earlier that they were going to meet us after they completed their workday. Karpenko sat in on all of the discussions. Before the arrival of Stanyukovich, I began to ask Pobedonostev questions.

Question. Is there a special space flight program set up for the future?

Answer. Yes, there is a special program set up for the future of space flight. The return of a dog to earth would be the first step.

Question. Have you experimented with any manned or rocket-powered aircraft?

Answer. During the International Geophysical Year we are exploring the upper atmospheric regions. Our first step is to get back containers from the satellites with recording apparatus; then, we will try for the return of animal containers.

Comment: At this point, Pobedonostev opened his briefcase and whipped out a copy of Missiles and Rockets magazine. He explained to our interpreter that he wanted to show us an article. This article appeared in the February 1958 issue of Missiles and Rockets, which was a special issue dedicated to electronics. The article had first appeared in an East German magazine called New Times. This magazine is also published in the Russian language and was picked up in a roundabout way by Missiles and Rockets from Novoye Vremya, which also means New Times, and was credited by Missiles and Rockets to a Russian magazine. At any rate, the theme of the article was that the United States is not very original. The United States manages to make progress only because of all the foreign scientists its brings to the States. Pobedonostev was perturbed by this.

Question. Why, in America, do you depend so heavily upon foreign scientists?

Answer. We do not depend that heavily upon foreign scientists.

Question. But for one thing you have many German scientists (rocket scientists), and also Timoshenko, father of modern structural mechanics went to America. He was a Russian, as well as the aircraft designers A. P. Seversky and I. I. Sikorsky, and a number of other experts.

Answer. These are only a small portion of all the scientists in America, many of whom are not very well known to the general public.

Comment: At this point I began to realize that I was being interviewed by Pobedonostev. A moment later Stanyukovich came in and was introduced all around; this temporarily broke up the interrogation. Pobedonostev was not to be put off, however, and he even dragged Stanyukovich back into this discussion of scientists from foreign countries being the mainstay of America's scientific endeavors. He mentioned Dr. Albert Einstein, who was a German, and also the Germans Von Braun and Dornberger, all taken to America. Frankly, it appeared to me that Pobedonostev, at least, had come prepared to be the interviewer and not the interviewed. His briefcase was full of Missiles and Rockets magazines to which he often referred. After the discussion of American science, he also leafed through pointing out various advertisements. He seemed particularly interested in this when I mentioned a particular question.

Question. Has any work been done in the Soviet Union on celestial navigation?

Answer. Some work in this field has been done in direction and speed and coordination, but it is all theoretical so far.

Question. Don't you feel that if you were to send men into outer space in a space ship that a very advanced electronics navigation system will be required?

Comment: At this point, Pobedonostev whipped out Missiles and Rockets again, leafed through it and found an ad about inertial navigation. He pointed to the ad and repeated that the work in the Soviet Union is all theoretical so far.

Question. Is there a special course in astronautics? Are students being trained in specializations directly connected with space flight?

Answer (Stanyukovich). No.

Question. When will a research vehicle be sent to the moon? Is it being worked on actively?

Answer (Stanyukovich). With the status of U.S. technology and the Soviet system of incentive and enthusiasm, we could have men on the moon in no time.

Question. How would you go about bringing together the United States and the Soviet Union in such a cooperative effort?

Answer (Stanyukovich). It is up to the United States. We are willing to cooperate.

Question. But in what specific way could we go about achieving this kind of cooperation which I feel would be of tremendous help toward easing international conscience?

Answer (Stanyukovich). There is too much secrecy in the United States and also you continue to send Strategic Air Command bombers toward the borders of the Soviet Union. This is no way to build good will or trust.

Question. But does not the Soviet Union also have security measures?

Answer (Stanyukovich). Not at all.

Question. You mean that everybody knows what is going on at all times?

Comment: There was no one answer at this point. All three, Karpenko, Pobodonostev, and Stanyukovich, emphatically claimed that security does not exist in the Soviet Union. There is no need for it. Everybody is working for the common good. There is no rivalry. Two hundred million people own the state. Scientists, engineers, and workers all hold conferences, argue out problems, and work together. These three academicians claimed they had not seen the sputnik firings.

Question. Do you know about the construction, configuration, and method of the engines used of the final successful launching of sputniks?

Answer (unanimously). That is something the engineers know.

Question. But have you no idea of the type of rocket engines used to launch the sputniks? Were these engines single engines, engines in cluster, or were the rockets multistage rockets?

Answer (Pobedonostev). All kind sof propulsion systems were tried. Single engines, engines in cluster, multistaged rockets, and so forth.

Question. But you do not know which of these was most successful and finally used?

Answer (Stanyukovich). That is a problem for the engineers.

Question. Has any research work or actual experimental work been done on full pressure suits?

Answer (Stanyukovich). We have pressure suits.

Question. No, I mean full pressure suits, not partial pressure suits. The type of suit that would be operative or effective in outer space.

Answer (Dobedonostev). We will develop those types of suits when we need them.

Question. But how will you send men out into space without full pressure suits?

Answer (Pobedonostev). Dogs will go first into space.

Question. But in the United States before we came to the Soviet Union I read many front page stories in the newspapers about the Soviet Union training a group of men for a flight to the moon. Could you give me some details about this?

Answer (Pobedonostev; he laughs). The whole thing began with a newspaper story. There was an Italian newspaper reporter here visiting Moscow from the newspaper Lunita. This reporter tuned in a radio and heared a children's science fiction story. He thought it was a true report since he had tuned into the program in the middle of the story. He then went back to Italy and published it as a factual story. I myself saw the Italian newspaper, and it was a front page story. [He laughs again.] It was picked up by the French newspaper L'Humanite where somehow a picture of one of our pilots was shown. The caption said that the pilot was launched in a rocket. [All three men laughed at this point.]

Comment: Stanyukovich then told another story to illustrate the point of how newspapers distort the truth internationally. Sanyukovich's story went as follows: In the German People's Republic, that is East Germany, a paper published a story last April 1 about five persons already launched in rockets. The newspaper reporters were influenced namely by their salaries. They felt that they had to work a spectacular story to keep their jobs. This is true all over the world.

Question. Your sputniks are inclined 65° to the Equator which makes their orbits almost polar. How were you able to achieve this?

Answer (Stanyukovich). Just by using a little more power and a little more speed.

Question. Could you be a little more specific?

Answer. We are theoreticians—not engineers.

Question. At the Moscow planetarium not long ago I saw a motion picture which included a diagram of a sputnik launching rocket. It showed a three-stage rocket with a group of rocket motors in the first two stages. I believe I counted five in the first stage and three in the second. The third was a single engine rocket. Could you tell me a little bit more about this?

Answer (Pobedonestev and Stanyukovich). This was for purposes of explanation at the lecture. Those pictures and diagrams were not of the actual rockets used.

Question. Do you have an absolute program with definite dates for the launching of sputniks?

Answer (Stanyukovich). Yes, we do.

Question. How closely to these dates to you come in your launchings?

Answer. We were 1 year ahead on Sputnik I.

Question. How do you explain your ability to beat your own schedule by such a great length of time?

Answer (Stanyukovich). It is because of the way things are organized and the great enthusiasm of the workers who went all out to put their efforts into this program. Our original plan was to launch the first sputnik at the end of the International Geophysical Year, but instead we got it up there first.

Question. How many people were involved in the successful launching of the first sputnik?

Answer (Stanyukovich). 170 million people were involved.

Question. But I mean specifically what kind of people; how many propulsion engineers; how many electronic specialists; how many airframe designers?

Answer (Stanyukovich). It is very difficult. The rockets needed to launch the sputnik are very difficult. One group of people works on the airframe, another group works on the engines; another on the navigation systems, a fourth group on the measurements, the recording apparatus. The fifth group studies the medical problems. This with the second sputnik and the dog Laika. All groups are helped by the metallurgist, thermodynamicists, astrophysicists who are working on cosmic radiation measuring apparatus, and so forth. For example: There was one big group of people working on just transistors alone. The most qualified people in each field worked on the tasks of the sputniks to get the answers for the problems and to meet the schedule.

Question. In other words, these people were coordinated in one vast effort to launch the sputnik?

Answer (Stanyukovich). No, these people worked at the regular electronics and engineering factories.

Question: You mean that these were the regular factory and institute workers that produced the components for the launching of the sputnik as well as the sputnik itself?

Answer (Stanyukovich). Yes.

Question. Didn't this interfere with regular production in those factories?

Answer (Stanyukovich and Pobedonostev). Certainly not.

Question. It would seem to me that if everybody were concentrating on this one project there would be little time left to do their regular work?

Answer (Stanyukovich). Our workers were so enthusiastic when they learned about the project that they were working on that they devoted their spare time to working on the sputniks. It did not interfere with regular production.

Question. Was there any one person who directed or carried out the sputnik program?

Answer (Stanyukovich). There was much exchange of opinions and results between the scientists. From this we got the best information for the production of the sputnik. The heads of different Soviets coordinate work in each field then give the task to each particular plant.

Question. What do you mean by Soviets?

Answer (Stanyukovich). Each Soviet is a complete specialized society. By having all these technological societies coordinated by a greater society they are able to accomplish what normally might seem impossible.

Comment: All through this interview I was trying to find out about manned rocket-powered aircraft such as the American Bell X-1, X-2, and the X-15. And always when I mentioned it, they would deviate and change the subject.

Question. Are Soviet scientists working definitely toward manned space flight?

Answer (Pobedonostev). Yes, we are working toward manned space flight. We started this program with the dog Laika in Sputnik II. We had very many complaints from all over the world about Sputnik II and the dog. From the United States we received a cable from a woman in Detroit, Mich. She asked us urgently to save the dog's life.

Question. Well if you have a definite program toward manned space flight then you must have rocket-powered aircraft?

Answer (Pobedonostev). We have jet aircraft.

Question. No, I know you have jet aircraft. I mean experimental pure rocket-powered aircraft.

Answer (Pobedonostev). In 1937 I was the first pilot to fly in a rocket-powered aircraft.

Question. You flew in a rocket-powered aircraft?

Answer (Pobedonostev). Yes.

Question. But that was in 1937, I didn't realize there was rocket-powered aircraft then?

Answer (Pobedonostev). This was a small light airplane.

Question. And you flew it on pure rocket power?

Answer (Pobedonostev). It was a liquid rocket.

Question. But was it purely rocket powered?

Answer (Pobedonostev). It was a rocket-assisted aircraft.

Question. No, I do not mean rocket assisted; I meant pure rocket powered such as the Bell X-1 series, the X-2, and aircraft like the proposed X-15 in the United States.

Answer (Pobedonostev). (He pauses, looks a little bewildered, looks at Stanyukovich who looks at him and then he shakes his head no.)

Question. Has the Soviet Union ever experimented with rocket-powered aircraft such as the Bell X-1 and X-2 series?

Answer (Stanyukovich). We do not believe in such aircraft.

Question. But it is precisely this kind of aircraft that trains pilots in the rigors of extremely high altitude flights and also, therefore, space flights?

Answer (Pobedonostev). Such kind of flying is much too risky.

Question. It is a calculated risk. There are many pilots in the United States who have flown rocket-powered aircraft and for a manned space flight program such aircraft are, I feel, very necessary?

Answer (Stanyukovich). Human life is very sacred. We always consider the safety of the pilots.

Question. But the safety of the pilot is always considered in the United States too, at Edwards Air Force Base, Calif., where these rocket-powered aircraft are flight tested. Every precaution is taken to assure the safety of the pilots.

Answer (Pobedonostev). They are very risky aircraft.

Question. But has the Soviet Union ever experimented with such aircraft?

Comment: At this point Stanyukovich gets very angry—obviously angry—and starts a long harangue in Russian with Pobedonostev. Karpenko, who has been sitting in, more or less as moderator on this whole interview places his hand on the arm of Stanyukovich to calm him down. Pobedonostev is slightly perturbed too, but not nearly so much as Stanyukovich. Our interpreter doesn't translate any of this byplay which goes on for at least 3 minutes, possibly longer.

Question. But I am only trying to find out if, in your space-flight program, you make use of pure rocket-powered experimental aircraft.

Answer (Pobedonostev, after a long pause). No, we have not.

Question. But you said that you had a definite program with definite dates.

Answer (Pobedonostev). We do.

Question. Then, since physiologically and psychologically, it is most important to know how a space pilot will act under space conditions, what are you doing in this respect if you don't have rocket-powered aircraft?

Comment: Stanukovich fidgets angrily again.

Answer (Pobedonostev). We make use of dogs in rockets and sputniks.

Question. But a dog is hardly a substitute for a man in terms of consciousness.

Answer (Pobedonostev). The dogs will lead to the time when we can use men. One cannot know where he is going until he gets there.

Question. If you don't know where you are going until you get there, then you must take risks. You claimed that rocket-powered aircraft are too risky for human life, but aren't such risks necessary for the successful achievement of space flight?

Comment: At this point Stanyukovich practically explodes in anger, and Pobedonostev, himself, who is normally a phlegmatic type gets flustered and also angry. Karpenko also again has to act as the pacifier for approximately (I would say) 2 or 3 minutes. There is a big argument between all three with Karpenko calming both men down.

Question. I think we all agree that the present types of chemical-powered rockets are quite awkward and primitive, to really explore outer space, not only within the solar system but among the stars requires new propulsion systems. In fact even to get to the moon nuclear-powered rockets would be much more efficient than a chemical-powered one. Does your program in the Soviet Union include research and experimental work in propulsion methods or systems such as the nuclear, the ion drive, and the photon drive type of propulsion?

Answer (Stanyukovich). Some work has been done in the ion drive.

Question. You mean experimental work?

Answer (Stanyukovich). It is theoretical work.

Question. And in the fields of nuclear-powered rockets, has any experimental work been done?

Comment: There was a pause, and no direct answer. Stanyukovich implied that work was being done on nuclear-powered aircraft, but completely skirted the question which dealt with nuclear-powered rockets. I forget his exact phraseology but this was the general effect. I then asked him the following question:

Question. In the United States there are companies like Glenn L. Martin who have set up a research laboratory to investigate the use of gravity itself as a means of propulsion. The so-called antigravity drive—is there any such investigations being considered as part of your space-flight program?

Answer (Stanyukovich vehemently). Martin deceives people. The gravity drive is impossible now. It is a very fashionable thing. One out of every ten people are interested in it now and I think it is much better to use the quanta theory, using a gravitational relationship of particle field; the magnetic fields of particles. I feel that this could be accomplished within a minimum of 10 years.

Question. But what sort of propulsion system is this? What do you mean by using the magnetic fields of particles, and how would you use them to drive a spaceship?

Answer (Stanyukovich). I do not know the details; however, Prof. Dmitri Dmitrich Ivanenko has been working this theoretically: If you would get in touch with him, he would be able to give you more details.

Comment: To close the interview I reiterated the fact to Stanyukovich that new types of propulsion methods would be necessary in order to explore the universe. For the first time in many minutes he smiled and agreed and he made the following quote: "Yes, the earth is now thoroughly explored. The geography of the earth, we can say, is school science. Perhaps within a hundred years the geography of the solar system will also be called school science." Throughout this interview which lasted approximately 3 hours, all three of the Soviet scientists continually glanced at their wristwatches but were very polite and patient in answering questions except for the several outbursts of anger on the part of Stanyukovich, I had the impression that they were quite sincere in their answers but either they didn't know everything I had assumed they knew or else they skillfully avoided answering certain questions. The anger on the part of Stanyukovich, I believe, was stimulated by my insistence on learning about rocket-powered aircraft. The interview was concluded after considerable small talk following the last question, and once again I became the person interviewed as they asked many things about U.S. education and scientific specialization. Strangely, Pobedonostev seemed to think that one of the big faults of the United States was too much specialization and not enough broad knowledge on the part of the scientists. I pointed out to him that although this was true of many scientists, not only in the United States but throughout the world that in the United States we have men like Dr. Fritz Swicky and many others whose attitude was that all the sciences have an interconnection and that all scientists should be familiar as much as possible with all sciences in order to be most efficient in their own field. Equally strange was the fact that during our thousands of miles journey through the Soviet Union, no other Soviet scientist ever, with two exceptions, seemed to agree with Podebonostev's attitude. They were all for specialization, particularly in engineering.

INTERVIEW IN MOSCOW WITH ANATOLY GRIGORYEVICH KARPENKO, SCIENTIFIC SECRETARY OF THE PERMANENT COMMISSION ON INTERPLANETARY COMMUNICATIONS OF THE ACADEMY OF SCIENCES, U.S.S.R.

(Interview by Lloyd Mallan, American author)

Comment: This meeting took place in the Intourist Service Bureau of the Savoy Hotel, where we were staying in Moscow. Mr. Karpenko did not, at that time, invite us to the Academy of Sciences, although originally that is where we thought we would interview him. Before asking him any questions, I mentioned, in passing, that I thought Dr. Fred Singer, who is a well-known American physicist at the University of Maryland, and the man who first conceived of a practical minimum earth satellite, deserved to be awarded the new Russian Theocoskiy gold medal. The Russian Academy of Sciences, U.S.S.R., dreamed up this award, which not only included the gold medal but also a small sum of money for outstanding contributions to space flight on an international scale. I really believed, and I still do, that of all the people who deserve such an award, Dr. Singer does. Mr. Karpenko nodded vehemently when I suggested this. I might add that nothing has ever been done about it. This meeting took place about the third week of April 1958.

Question. What is your attitude about Dr. Eugene Säenger's photon-driven rocket?

Answer (a laugh). Dr. Säenger is a very great scientist. He is of the greatest in the physics of propulsion, but the photon rocket we will never see in this country; perhaps a hundred years from now, yes. You know this is my private opinion, of course. But in the very close future, 10 to 20 years from now, I believe the ion rocket will permit man to visit other planets of the solar system.

Question. Why are you so enthusiastic about the ion rocket and so sure that it will be feasible in so comparatively short a time and do not think the proton rocket will be something feasible for many, many decades?

Answer. The ion propulsion method is a pet of mine; in fact, I am doing considerable research on this right now. I know it is feasible.

Question. What would you say would be the maximum time from today that such a rocket of the ion type could be feasible?

Answer. I will say that the maximum time from now will be no more than 20 years.

Question. Is there any experimental work being done on the ion rocket at this moment in the Soviet Union?

Answer. No; everything is theoretical right now.

Question. Why, then, if work is only theoretical on the ion rocket in the Soviet Union, and also the photon rocket is a theoretical thing as well, and Dr. Eugene Säenger is energetically doing theoretical work on it in Stuttgart, Germany, can you be so certain that in two decades the ion rocket will be feasible and the proton rocket will not?

Answer. Although no experimental work is being done now on the ion rocket, such work could begin at any time if the advancement of theory warrants it. All we would have to do would be request a special appropriation for funds.

Question. Fine; but why wouldn't this also apply to the photon rocket?

Answer. At present only theoretical work is possible in this field. There is no possibility for practical work. To even discuss the photon rocket today would be like discussing our present working rockets in a technological civilization like that of ancient Egypt.

Question. Well then, is there any work actual or theoretical going on among Soviet scientists as regards the photon rocket.

Answer. No.

Question. Regarding pure research types of rockets could you tell me just how many and what kinds have been developed and are being used in the Soviet Union.

Answer. There are two types of pure research rockets in the U.S.S.R.; one is a small meteorological probe rocket, somewhat like the Aerobee configuration but launched through a portable launching tower. This is used to investigate the upper stratosphere, winds at high altitudes, chemical composition of the stratosphere, and so forth. The other is a very big rocket for ramal, meteorological, and geophysical research into the ionisphere. It can carry a payload of 1½ tons.

Question. Are there any rocket-powered research aircraft in the Soviet Union such as the Bell X–1 and X–2 series? If so, what experiments have been carried out with them?

Answer. I have never heard of any rocket-powered aircraft in the Soviet Union. I cannot recall any experimental craft like these in use.

Question. But in terms of space medical research, don't you think that pure rocket-type aircraft such as these that I have just asked you about are invaluable for space medical research?

Answer. Your Dr. David Simon's experiments with stratospheric balloons are much more valuable to space medical research. Simons is a very brave man. He was up where cosmic radiation might do him harm for a very long time. Nobody knows now whether damage was done to him or not. It may not reveal itself for many years.

Comment: Karpenko's attitude about David Simons was derived from the paper that Simons gave at the International Congress of the International Astronautical Federation at Barcelona in 1957.

Question. Has the Soviet Union or the Academy of Sciences of the Soviet Union ever carried out a series of high altitude balloon experiments such as those accomplished by Dr. Simon?

Answer. There are none.

Question. Is there any work experimental or otherwise planned?

Answer. No.

Question. Well then, how do you propose to design space suits? Designing such suits for space-going pilots requires a background of experience in high altitude physiology?

Answer. When the time comes to send men into cosmic space we will construct those suits. Then we will call on our background of knowledge and put them together when we need them.

Question. But how will you acquire this background of knowledge?

Answer. From the dogs that we send into space.

Question. Are there space suits designed for dogs?

Answer. No, there are no space suits designed for dogs. We pressurize the dogs in their cabins.

Question. You have mentioned that you are acquiring your background toward the development of space suits through the use of dogs? Is there any special reason why you prefer dogs? Is it because, for example, I know that their respiratory systems are the closest to the human among mammals?

Answer. No, it is not because their respiratory systems are very human that they are used in our experiments. We use them because they have a very high order of nervous system; they respond much more easily to training; they can be taught to be content to unusual environment. In the United States your scientists like to use monkeys. Physiologically they are much similar to humans.

48438 O—59——2

But because of this they are very hard to train to be satisfied so you must anesthetize them. There is an old Russian proverb which goes, "He who is soused cannot answer questions." Or another translation, "A drunken man will give drunken answers." This would be true of drugged monkeys, but alert dogs can supply many answers about how the nervous system operates in distress of rocket flight and orbital lightlessness. We are very much interested in those answers about the nervous system which includes emotions as well as theological reaction.

Question. Do you plan to ever use any other mammals other than dogs for future experiments?

Comment: I asked this question because at one time after the launching of Sputnik II, the American press made a big issue about the fact that the Russians would next launch an ape in satellite.

Answer. No, we do not plan to use mammals other than dogs for future experiments.

Question. Well what exactly, in terms of psysiological reactions, do you intend to learn from dogs?

Asnwer. Many things are learned about fear for example: this again requires conscious mammals. A few days after they return from the rocket flight they are led into their chamber or a similar one. By the way they react for or against being led into this chamber we can find out how permanent their original conditioning is and how strong were the negative emotions, or the negative emotional conditions. Also how much they were conditioned spontaneously by the forces of the flight itself.

Question. I have read in many places that the Soviet Union was preparing to send an ape into orbit in sputnik III.

Answer. This was never even considered here.

Question. Well, then, I also have read in a number of reliable American newspapers that the Aeronautics Research Institute here in Moscow was in the process of training a group of men, in other words a crew, for the first manned flight to the moon?

Answer. If we were training such men now, by the time we would have them ready they would be much too old to take advantage of their training.

Question. The press, in this respect also in the United States, for a number of days after you launched sputnik II with the dog Laika, spoke of a recovery system when the time came for catapulting Laika out of the satellite and bringing her back to the earth alive?

Answer (a laugh). This whole story even fooled our own press because they also published the same story.

Comment: The assistant director of the Moscow Planetarium gave a lecture on the orbits of sputnik. After he was finished, a newspaperman of our country asked the lecturer, who by no means was a scientist or a member of the Commission on Interplanetary Communications, whether or not the dog would be brought back to earth alive. The kindhearted lecturer answered, "Yes, I should hope so," or words to that effect. It was a personal wish on his part and either through mistranslation or misquotation it was taken up by the world press including the Soviet Union.

Question. Well at any rate has any practical work been done on escape devices or escape capsules, or have any of these been contemplated for the sputniks.

Answer. Such devices are being worked on but the problem is far from being solved as yet. When they are solved manned space flight or space flight by man will be possible.

Comment: Of most of the scientists we interviewed, Karpenko seemed absolutely sincere and spontaneous in all of his answers. There was no hesitation and, of course, no consultation with any other scientists before he answered.

Question. Why is it that work is going so slow in the development of an escape capsule to recover experimental animals alive?

Answer. Such a capsule might easily be greater in weight than the total weight of the satellite itself. Much greater thrust will be required of the rocket to get it aloft, and such thrust is not yet available.

Question. When do you think such thrust may be available?

Answer. Perhaps within 3 to 5 years.

Question. Speaking of the next 3 to 5 years what do you think are more than just casual possibilities?

Answer. The following, I believe, are quite possible within the next 3 to 5 years:

1. Recovery of mammals from satellites.
2. Recovery of the satellites themselves.
3. A rocket to pass within thirty-odd thousand miles along the dark side of the moon.
4. A circumlunar rocket to orbit about the moon for several days.
5. An earthlunar sputnik to orbit between the earth and the moon.
6. Automatic tankettes on the moon's surface (remote controlled crawler-type vehicles with TV sets).

Question. You are absolutely sure that these are the logical possibilities or the practical possibilities, lets say, within the next 3 to 5 years?

Answer. It is expected that decisive solutions will be achieved in all of these fields to from 3 to 5 years.

Question. To continue in another field, in order to achieve a rocket orbiting about the moon or to send a rocket to the moon the most efficient way to do this would be a nuclear-powered rocket, would it not?

Answer. Certainly.

Question. How much research or practical experimental work has been done in the Soviet Union on nuclear-powered rockets?

Comment: There was no answer here. Karpenko shrugged almost imperceptively and began to talk about ion-propelled rockets, which are distinctly different. I couldn't bring him back to the original question. I brought the question of nuclear rockets up as clearly as I could, twice. In both cases Karpenko evaded the issue by talking about ion-powered rockets. I don't think he could have misunderstood the word nuclear. Here are some closing comments that may be of some interest: One of them is with Proffessor Sedov, we mentioned to Karpenko that our time was short. We didn't know whether or not we could have our visa renewed and it was imperative that we see as many scientists as possible in the general field of space flight before we left the Soviet Union. Karpenko, at this point, suggested Federnov (a famous name in the Soviet Union, by the way). Federov is head or chief of the Propulsion Institute, of the Academy of Sciences of the U.S.S.R. Karpenko though, unlike Sedov and others we later met in the Soviet Union, immediately went to the telephone and tried to phone Federov so that we could see him the following day. This was 2 days before May Day, which is one of the biggest holidays in the Soviet Union. But, nevertheless, Karpenko felt that we could see this propulison expert. He spoke with Federov's secretary who told him that Federov had some hours earlier left Moscow for his summer home for the holidays. However, the secretary said that Federov would return to Moscow immediately after May Day. Karpenko then told us, after explaining the situation, that he would get in touch with Federov immediately after May Day to arrange a meeting. He never arranged a meeting with Federov but I am sure that he tried to get in touch with him to do so. Our purpose in this meeting was to shoot pictures at the Propulsion Institute. Since Federov was the head, he was the only man who could give permission to shoot those pictures. However Karpenko did subsequently arrange for us to see the head of the aviation industry and a top propulsion man, Stanyukovich, but not Federov.

Final human interest comment: Karpenko was not the least bit self-conscious except during the first moment of introduction. After that he was relaxed, voluble, and enthusiastic. He never hesitated to take offered cigarettes as other scientists did. He smokes as much as I do and once absently withdrew a cigarette from the pack that I had on the table, lit it, and puffed away while talking without waiting to have it offered to him. He was that absorbed. He seemed more businesslike, too, than the others.

Interview in Moscow With Leonid Sedov, Head of the Permanent Commission on Interplanetary Communications of the Academy of Sciences, U.S.S.R.

(Interview by Lloyd Mallan, American author)

Comment: We met Professor Sedov in mid-April in the evening. (The next portion of the tape was accidently erased). Mr. Mallan goes on to say that he met him in an institute of learning of some type within the city of Moscow. The first question I asked of Professor Sedov was about the status of astronomy and astrophysics in the Soviet Union. His answer: The best instruments are in the United States of America.

Question. How about radio telescopes?

Answer. Those too. The biggest and best instruments, both radio and optical, are in your country.

Question. Professor Sedov, we need your help. This book that we are working on, we feel that it is exceptionally important to international understanding and good will. Would it be possible for you to arrange for us to visit a research missile test center?

Answer. No problem [through the interpreter].

Question. Would it also be possible for us to make photographs of research rocket vehicles. Not necessarily military but ones that would be used in pure research?

Answer. No problem.

Question. Would it be possible to interview people working on these projects, because one of our points is to humanize science in Russia to the American public?

Answer. Of course. [Question by Professor Sedov]. What sort of people would you like to interview?

Answer (Mr. Mallan). Astrophysicists, astronomers, engineers in the fields of propulsion and airframe design, space biologists and other aeromedical research scientists, and electronic scientists as well as electronic engineers. That is, theoreticians as low as practical designers in the field.

Answer. We can arrange this. No problem.

Comment: Professor Sedov very carefully noted on a little pad that he carried each of the types of scientists and engineers as well as physiologists that we wanted to see. He not only carefully made the list, but he added a note when we mentioned that of all these fields to us the field of aviation medicine was of prime importance and in every case his answer was—no problem; that he would arrange it.

Question. We would also like to take photographs at the Institutes of the Academy of Sciences such as the Institutes connected with the subjects we have mentioned—like aeromedical research center, propulsion research center, and the aviation institute.

Answer. This can be arranged

Comment: As Professor Sedov made the list (I used the word designer) he said, "Oh, yes, you would like also to talk with one of our engineers." He mentioned this because I had used the word scientific worker to include both engineers and theoreticians in the general field of space flight.

Question. Professor Sedov, we have met Dr. Eugene Säenger in Stuttgart, Germany, who is working on proton propulsion for rockets. What is your attitude toward proton propulsion?

Answer. The proton rocket is a fiction of science. There is no foundation of fact to it.

Question. But, Professor Sedov, isn't it true that all great achievements in science are the result of visionary scientists who, ahead of their time, dove into areas like that of Dr. Säenger's?

Answer. Actually the answer here was a kind of shrug. I persisted.

Question. But would you say that the proton rocket is completely impossible?

Answer. Perhaps in 50 years it may be possible. There are many things, all complicated, that must be accomplished first. There will first have to be atomic electrical plants all over the world. And, of course, Dr. Säenger believes the proton rocket is possible. He has to believe it.

Question. Professor Sedov, could you tell us anything about your forthcoming sputnik launching, which will be No. 3?

Answer. Nothing will be known about that or other sputniks until they are working. It is of no interest to talk about them until they are successful.

Question. But today the sputnik program is the most interesting part of your space-flight endeavors, is it not?

Answer. Yes.

Question. Well, then, couldn't you at least extrapolate into the future.

Answer. A shrug of the shoulder.

Comment: The fact that many trade and technical magazines, as well as reliable newspapers in the United States such as the New York Times had mentioned that the sputnik program in the Soviet Union had been instigated by the Chkalov Aero Club. The Central Aero Club of the Soviet Union had proposed these earth satellites.

Question. How did it happen that the Chkalov Aero Club (the Chkalov Central Aero Club of the U.S.S.R.) began their promulgation for a program to launch earth satellite vehicles?

Answer. The Chkalov Aero Club did not instigate the space-flight program.

Question. Who did, then?

Answer. The Academy of Sciences of the U.S.S.R.

Comment: In conclusion I would like to summarize the interview with Professor Sedov. The entire interview lasted about 2½, perhaps 3, hours. In the process there was much small talk, of course, such as Sedov making quite a big issue of the fact that the Academy of Sciences of the U.S.S.R., after the launching of Sputnik I, had received thousands upon thousands of letters, as he put it, from people all over the world requesting that they be granted the right to act as volunteers to be sent up into orbit in the next sputnik, which would be Sputnik II. But the most important thing of all to us was the fact that Sedov implied—not exactly—he stated. It wasn't merely an implication. When we asked him to dinner that evening, he said that he had other things to do, but that within a few days he would be very happy to have dinner with us. The thing that was implied in other statements he made before we left him, was that we would be seeing him very often and that he would work with us to the best of his ability to help us achieve a great humanized picture of Soviet science, in terms of bringing out international good will and better understanding. Professor Sedov took from us a list of 60 Soviet scientists that we had prepared over a period of months in much research. He checked off the names of the scientists on that list who would be most pertinent to our project which was a project devoted to the field of space flight, which included all allied fields: Electronics, guidance, missiles, rockets—the whole business. And of the 60 he checked off approximately 35 names (as I recall now) and said that he would get in touch with all of these people. We never saw Professor Sedov again nor did we ever hear from him again by note or telephone; however, we spent approximately 10 days trying to get in touch with Professor Sedov ourselves. First using ordinary channels (that is, using our intourist guide and interpreter) and then even using devious means. We had met some Russian students who spoke English. One of them phoned a number of times for us representing himself as a Government official. At no time were we able to reach Sedov; although every time at his three offices, secretaries there took notes and later when we phoned back said they had delivered those notes. There were two exceptions to the list of about 35. Sedov did get in touch with Alla Masevich, to whom we had a letter of introduction which we had mailed to her at the same time we had mailed our letter of introduction to Sedov. She had never phoned us until the day after we spoke with Sedov, so I assumed that he had spoken with her as he said he would. The other exception was Karpenko, scientific secretary of the Permanent Commission on Interplanetary Communications of the Academy of Sciences. Karpenko phoned us saying that he would see us approximately 10 days after we saw Sedov.

MEMO

To: Committee on Science and Astronautics.
From: Lloyd Mallan.
Subject: List of scientists and VIP's who gave me letters of introduction to their colleagues in the Soviet Union.

1. Dr. S. Fred Singer, associate professor of physics, University of Maryland (to Prof. Leonid Sedov, head of the Permanent Interdepartmental Commission on Interplanetary Communications, U.S.S.R. Academy of Sciences).

2. Dr. John W. Evans, superintendent, Sacramento Peak Observatory of the Geophysical Research Directorate, Air Force Cambridge Research Center (general letter to Soviet Academy).

3. Dr. Richard E. Shope, of the Rockefeller Institute for Medical Research (to Prof. Viktor M. Zhdanov, U.S.S.R. Deputy Minister of Health and a specialist on viruses).

4. Dr. Armand N. Spitz, inventor of the Spitz Planetariums and head of U.S. Moonwatch program (to "whom it may concern").

5. James J. Harford, executive secretary of American Rocket Society (to Prof. Yevgeny Federov, chairman, Soviet IGY Committee for Rocket and Satellite Research).

6. Mr. Andrew G. Haley, general counsel of American Rocket Society and at the time president of the International Astronautical Federation, of which the Soviet Union is a member nation (24 letters of introduction, including the president of the U.S.S.R. Academy of Sciences, Nesmeyanov, the presidents of the Academy of Sciences of the Kazakh S.S.R. and Ukrainian S.S.R., as well as Mr. Khrushchev).

7. Dr. Eugen Sänger, director of the Forschungsinstitut Für Physik Der Strahlantriebe at Stuttgart, West Germany, and a world-renewned authority on rocket propulsion (to Professor Sedov, mentioned above).

NOTE.—Altogether, as I recall, I carried with me to the Soviet Union about 33 letters of introduction, some of the above persons having given me both specific and general letters of introduction.

Mr. ANFUSO. Mr. Chairman.

The CHAIRMAN. Mr. Anfuso.

Mr. ANFUSO. May I add to that, Mr. Mallan? Would you mind putting into the record the names of the Russian scientists you interviewed?

Mr. MALLAN. I have a typewritten list of 38 of those scientists I interviewed.

Mr. ANFUSO. Will you put that in the record, also?

Mr. MALLAN. I certainly will.

(The list of Russian scientists is as follows:)

LLOYD MALLAN INTERVIEWS—LIST OF SOVIET SCIENTISTS

MOSCOW

Leading scientists

Prof. Leonid Sedov, member of the U.S.S.R. Academy of Sciences and chairman of the Permanent Interdepartmental Commission on Interplanetary Communications of the Academy.

Prof. Alla Masevich, assistant director of the Astronomical Council of the Academy and head of the sputnik optical tracking program.

Anatoly Grigoryevich Karpenko, scientific secretary of the Commission on Interplanetary Communications of the Academy.

Yuri Pobedonostsev, professor of sciences at the Moscow Aviation Institute and member of the Academy of Sciences.

Kirill Stanyukovich, professor of gas dynamics and chairman of the mathematics department, Moscow Technological Institute, also member of the Academy.

Other scientists

Lev Viktorovich Zhigarev, expert in radio location techniques and coordinating editor of the popular science magazine, Znaniye-Sila (Knowledge is Power).

LENINGRAD

Leading scientists

Prof. Alexander A. Mikhailov, director of Pulkovo Observatory, corresponding member, Academy of Sciences of the U.S.S.R., and chairman of the Academy's Astronomical Council.

Prof. Gleb Chebotarev, director of the library, Leningrad Academy of Sciences, corresponding member of the Academy of Sciences of the U.S.S.R., and the scientist who calculated the trajectories of rocket for Project Boomerang, the Russian plan to send a sputnik around the moon.

Other scientists

Vladislav Sobolev, solar astronomer at Pulkovo Observatory.

Alexei Kisselev, astronomer at Pulkovo Observatory.

Andrei Nemiro, astronomer at Pulkovo Observatory.

Dmitri Shchegolev, astronomer in charge of sputnik tracking at Pulkovo Observatory.

ODESSA

Leading scientists

Prof. Vladimir Platonovich Tsesevich, director of Mechnikov Observatory and corresponding member, Academy of Sciences of the U.S.S.R.

Other scientists

Yuri Rousseau, astronomer at Mechnikov Observatory.

THE CRIMEA

Leading scientists

Prof. Andre Severny, director of the Crimean Astrophysical Observatory and corresponding member, Academy of Sciences, U.S.S.R.

Peter P. Dobronravin, assistant director, Crimean Astrophysical Observatory.

E. R. Mustel, head of the IGY Committee for the U.S.S.R.

V. B. Nikonov, stellar physicist, in charge of the stellar division at the Crimean Astrophysical Observatory.

Arnold Stepanyan, head of cosmic ray research at the Crimean Astrophysical Observatory.

Other scientists

L. S. Galkin, scientific secretary at the Crimean Astrophysical Observatory.

Nikolai Steshenko, solar physicist at the Crimean Astrophysical Observatory.

Benjamin Mojerin, head of sputnik tracking at the Crimean Astrophysical Observatory.

S. A. Savich, head of ionospheric and earth magnetism research at the Crimean Astrophysical Observatory.

ARMENIA S.S.R.

Leading scientists

Prof. Viktor Ambartsumyan, president of the Academy of Sciences of Soviet Armenia, director of Byurakan Astrophysical Observatory, and corresponding member of the Academy of Sciences of U.S.S.R.

Torgom Yessayan, assistant director, Armenian Academy of Sciences, and well-known philosopher.

Doctor of Sciences Ludvig Mirzoyan, scientific secretary of the Armenian Academy of Sciences.

Other scientists

Ramela Shahbazian, scientific worker at the Byurakan Astrophysical Observatory.

Harutjunyan, scientific worker at the Byurakan Astrophysical Observatory.

KAZAKHSTAN S.S.R.

Leading scientists

Dr. Kanysh I. Satpeyev, president, Academy of Sciences of the Kazakh S.S.R.

Academician Alexander P. Polosukhin, director of the Institute of Physiology of the Kazakh Academy.

Academician Vasily Fesenkov, director of the Astrophysical Observatory of the Kazakh S.S.R. and corresponding member of Academy of Sciences of the U.S.S.R.

Prof. Dmitri Rozkovsky, astrophysicist at the Kazakh Astrophysical Observatory.

Prof. Gavril Tikhov, head of the Department of Astrobotany of the Kazakh Academy and corresponding member of the Academy of Sciences of the U.S.S.R.

Dr. Mihad Karimov, assistant director of the Astrophysical Observatory of the Kabakh S.S.R.

Dr. Zoya Koragina, astrophysicist at the Astrophysical Observatory of the Kazakh Academy of Sciences.

Dr. Nikolai Divai, astrophysicist at the Astrophysical Observatory of the Kazakh S.S.R.

Other scientists

Eugene Federov, scientific worker on the sputnik tracking camera at the Kazakh Astrophysical Observatory.

Andrei Kharitonov, astrophysical scientific worker at the Kazakh Astrophysical Observatory.

Mr. OSMERS. Mr. Chairman.

The CHAIRMAN. Mr. Osmers.

Mr. OSMERS. How long were you in the Soviet Union on this mission?

Mr. MALLAN. Two months.

Mr. OSMERS. Thank you.

Mr. MALLAN. And I traveled 14,000 miles in the process during which I was able to notice a lot of things. If you are alert and looking for specific things, you can see things that the average person would not know was there.

Mr. TEAGUE. Why, Mr. Mallan, do you think they allowed you to travel those 14,000 miles?

Mr. MALLAN. It was not easy. They resisted me on the itinerary for about 3 weeks, but it was finally a bit of bureaucracy and luck on my part that turned the trick. If you want to hear the explanation, it will take a few minutes. But the reason for the distance and these various localities was that the head of the Soviet optical tracking setup, Prof. Alla Masevich was kind enough to phone and write personal letters to astrophysicists that she knew plus the fact that I had already paid in advance for an open voucher before going to the Soviet. The fact that Masevich had arranged for me to see certain astrophysicists plus the fact that the Soviet Intourist Bureau did not know quite how to cope with an open voucher finally solved the problem with me.

It was a combination of luck and stubbornness really.

Mr. FULTON. Considering your research, your trip to the Soviet Union, your consulting with Soviet scientists as well as other scientists, and the other information you have been able to get up to this date, would you give us your considered judgment whether the shot announced by the Russians on Friday, January 2, 1959, was or was not a hoax? If you will answer yes or no, please.

Mr. MALLAN. I am sorry. I did not quite follow that in terms of the positive and negative answer. I think it was a hoax. I know, in fact, as far as I know, it was a hoax.

Mr. FULTON. As a result of all your investigations as well as your trip to Russia, your consultations with Russian scientists, with American scientists and any other research you have done, at the present time you say definitely it is a hoax.

Mr. MALLAN. That is right; yes, sir.

Mr. FULTON. Thank you.

Mr. OSMERS. Mr. Chairman.

The CHAIRMAN. Mr. Osmers?

Mr. OSMERS. I wonder if the witness would tell us whether he was able to convince any American scientists who thought it was not a hoax. Were you able to convince them that it was a hoax after you had presented material similar to the material you are presenting here?

Mr. MALLAN. If, by scientists, you also include tracking experts—you know, technicians such as radar specialists and microwave specialists—I can answer that affirmatively; yes.

Mr. OSMERS. Well, since you have made the distinction, I will have to question about the distinction.

Where would you draw the line between the tracking technician and the scientist in this case?

Mr. MALLAN. Well, the difference is a formal one. A scientist has a degree; a tracking technician may know just as much and not have a degree.

Mr. OSMERS. Well, I will narrow the question down. Were you able to convince any of the scientists, any of those men that had degrees, that it was in fact a hoax when they originally believed otherwise?

Mr. MALLAN. I was able to have them reverse their judgment and admit that there was a reasonable possibility that it was a hoax.

Mr. OSMERS. You were able to?

Mr. MALLAN. Yes; I was.

Mr. ANFUSO. Mr. Chairman.

The CHAIRMAN. I promised to recognize Mr. Wolf here for a moment.

Mr. WOLF. Mr. Chairman.

Why do you think, Mr. Mallan, that because you were in Russia and because you visited with these particular technicians that they told you everything? Now I was in Russia. I traveled several thousand miles, too, and I came away in 1957 with the strong feeling that they had told me some things but they surely had not opened up everything for me.

Of course, I was not concerned with technical matters and I have no reason to believe that because you went there and because you visited with them, that they gave you everything.

Mr. MALLAN. Well, as an experienced science reporter, I asked them the kind of questions that they had to answer honestly. I did not, for instance, ask them, "Do you have an intercontinental ballistic missile?" Instead, I asked them for the focal length of their tracking cameras and the type of lens arrangements. I asked them about their electronic guidance systems and I also closely inspected the equipment in both of these fields that they very proudly showed me, and this equipment was primitive.

I took photographs of it and I have at least proof photographs of these along with me.

Mr. WOLF. Well, I had the feeling after my trip to Russia, that what they wanted me to do was to come home and lull people to sleep in this country with a false sense of security about our superiority in the technical field. I am wondering if this is not what they wanted you to come home with as a great technician, as a great scientist in this field.

Mr. MALLAN. I am not a great scientist. I am a science reporter.

Mr. WOLF. Well, you presented facts in your magazine that would be designed to——

Mr. MALLAN. A science reporter. And all reporters should be able to present facts, as these reporters see them.

Mr. WOLF. I will rephrase the statement and say you presented ideas of your own that will be accepted by many people in a sense of lulling the American people to sleep. This is the question I ask.

Mr. MALLAN. The article also included facts. For example, Prof. Alla Masevich, a very noted astrophysicist in the Soviet Union and, as I mentioned earlier, is chief of their whole sputnik optical tracking setup, took me up on the roof of the Sternberg Astronomical Institute, which is the astronomical section of the Academy of Sciences of the U.S.S.R., and showed me their tracking camera. She was extremely proud of it. It was a modified Red air force aerial camera, with a WAF 2.5 lens. It had a fairly wide angle. It was a decent camera but it was extremely "old hat" stuff and not at all efficient. She mentioned to me, when I asked her about a newer type of camera, that they were working on one and she told me where they were doing this work.

I later went to Leningrad at Pulkova Observatory and spoke with Prof. Alexander Mikhailov. He is in an even higher position with the Academy of Sciences of the U.S.S.R. since he is chairman of the Astronomical Council and director of Pulkova Observatory, which is the coordinating observatory for the entire Soviet Union.

He told me that, yes, Maksutov, their famous optical expert, has been working on an F-1 camera.

He said—and you will find it in the tape recorder transcript—because Professor Mikhailov spoke English and I recorded the conversation—he said, yes, we are working on an F-1 camera which is quite like the ones you have; you know. Our super Schmidt Baker Nunn camera that was built for our optical tracking system throughout the Western Hemisphere.

They kind of astounded me because two sputniks had already been launched and at that time I believe we had launched only one satellite, Explorer 1, successfully. We had this terrific, predesigned, advanced setup of a radio and optical tracking system interlocked, tied in with computing centers so that we were not only able to follow and forecast the positions of our own satellites, but we were the ones who gave the Soviets the orbits for their satellites.

This, in itself, when I learned it, was a shock to me. But I can understand why we had to give them the orbits after I saw their tracking equipment.

Professor Masevich told me that the equipment she showed me on the roof of the institute was used at every key satellite tracking station in the Soviet Union.

I visited six of these key stations, spread over a distance of 10,000 miles, I would say, and in every case they were using this same "old hat" obsolete type of aerial camera.

Mr. Anfuso. Mr. Chairman.

The Chairman. Mr. Anfuso.

Mr. Anfuso. Mr. Chairman, I think that in order that we can properly evaluate Mr. Mallan's testimony, we ought to first offer for the record his story "The Big Red Lie." I think that should be printed in the record. I think we ought to do that first, and then I have another question to ask.

The Chairman. If there is no objection, the story will be printed in the record at this point.

(The True magazine reprint of "The Big Red Lie" follows:)

TRUE
THE MAN'S MAGAZINE
35c MAY 1959
A FAWCETT PUBLICATION

The Big Red LIE!

Russia Tries Biggest Hoax In History!

The BIG RED LIE

Confidential Advance Material

Not For Release Or For Public Information

Until Monday, A. M., April 20, 1959

Approved For Release 2004/08/31 : CIA-RDP63T00245R000100290011-7

The BIG RED

TRUE Correspondent Lloyd Mallan (shown above in an interview with Prof. Alexander Mikhailov, leading Soviet astronomer) is a veteran science writer/reporter, the author of numerous scientific books. Armand Spitz, head of the U.S. Moonwatch program calls Mallan "a superb observer, analyst and interpreter. . ." Dr. Fred Singer, designer of the first feasible earth satellite, introduces Mallan as "a science writer of great talent." Andrew Haley, president of the International Astronautical Federation, says of Mallan: "He is one of the truly great researchers. . ."

By LLOYD MALLAN

On Friday, January 2, 1959, the Soviet Union announced that it had successfully launched a rocket toward the moon. The Russians also claimed that this rocket, which they dubbed "Lunik," subsequently went into orbit around the sun.

This was a monumental triumph of propaganda. In one shattering blow, it wiped out the effects on world opinion of three great American achievements: the two Pioneer rockets that had nosed farther into space than anything previously made by man, and the delicately instrumented Atlas satellite that had opened a new era of long-distance communications. The world was agog. U.S. scientists writhed in chagrin and admiration. President Eisenhower himself, in a heroic show of sportsmanship, offered the Kremlin his genuine congratulations.

As for me, I was stunned.

In those first wild hours after the Russian announcement, I walked around in a nightmare. I was faced with a clearcut choice between two fantastic propositions. Neither seemed acceptable, but somehow, I had to choose.

The first proposition was that something was wrong with *me*, that in some unfathomable way I had been led into a colossal error of judgment. The second proposition was, simply, that the Russian moon rocket did not exist.

A few months before, I had traveled 14,000 miles through Russia on a scientific reporting expedition. I had talked to 24 of the top Soviet scientists—far more, I'm told, than any other Western journalist has ever been lucky enough to reach. I had seen major Soviet universities, research centers, observatories. Seemingly eager to have me take home a glowing report of their progress, the Russians had proudly shown me the cream of their technological achievements. I had looked and listened carefully. I had formed conclusions, tested them and re-tested them.

And I had come out of Russia with the dead, flat certainty that the United States was immeasurably far ahead in space technology—that, for one thing, Russia had no effective intercontinental ballistic missile; and that, for another, the U.S. would beat Russia to the moon without half trying.

But here I stood on the evening of January 2, Russia's epic announcement ringing loudly and mockingly in my ears. I poured myself a drink and sat down. I combed back through the notes and tape recordings and photographs and memories of my Russian journey. But no matter how I sifted and re-evaluated, I couldn't make it add up. Every

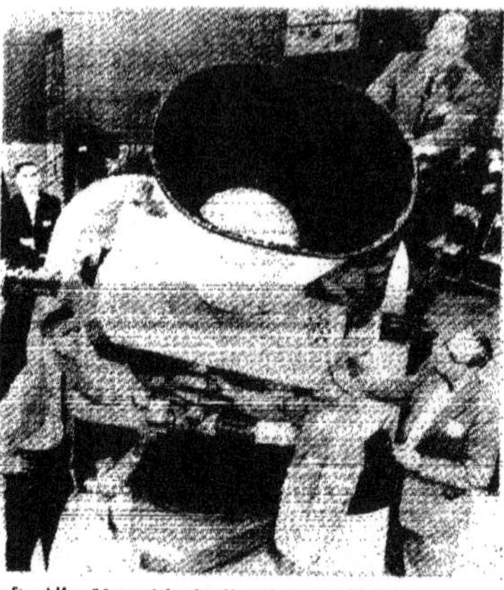

Left, Alla Massevich, leading lady of Soviet astronomy, watches an aide demonstrate best Russian aerial camera for Author Mallan. Compared to $100,000 U.S. tracking cameras (above—which can photograph a golf ball at 1,000 miles) the best Russian satellite tracking equipment appeared primitive.

This qualified observer, after 14,000 miles behind the Iron Curtain says:

1. The Soviet Union's first man-made planet, "Lunik," does not exist and never did.
2. The Russians do not have any ICBMs, the long-distance terror missile with which Khrushchev has threatened this country

thing I'd seen and heard in Russia argued against the alleged fact of Lunik. The scientific community which I had studied in that enigmatic land was not capable—simply not capable—of producing any such thing.

It was a hard concept to grasp, both intellectually and emotionally. But this was it: The Russians did not fire a rocket past the moon on January 2, 1959. If they fired anything, it failed to reach the distances achieved by the U.S. Pioneers. The Lunik, in short, was a coolly insolent, magnificent, international hoax.

I couldn't just let the incredible thought ferment in my brain, of course. I went to Washington, talked with military men and intelligence officers in the Pentagon. I visited Project Space Track, the Air Force installation in Massachusetts that collects and correlates tracking data from all over the Free World. I telephoned major tracking stations. I talked with scientists.

Not one of them would make the flat statement that he had heard a signal from Lunik. Officially, the U.S. was acknowledging the existence of Lunik. Unofficially and privately, the wet cold edge of doubt was beginning to seep into some clever minds.

Slowly, this doubt emerged into the open. Puzzled little essays began to appear in newspapers. Syndicated columnist Fulton Lewis, Jr., for instance, wrote on January 21 that "intelligence sources" were questioning Russia's veracity. An editor of the magazine *Electronic News* ran an exhaustive probe of the affair through correspondents around the world, ended convinced that no Lunik exists. He couldn't print his conviction, however; his publisher feared repercus-

PRESIDENT EISENHOWER:
"We seem very prone to give 100% credence to some statement of the Soviets, if it happens to touch upon our own anxieties. . . Apparently they are believed all around the world and too implicitly."

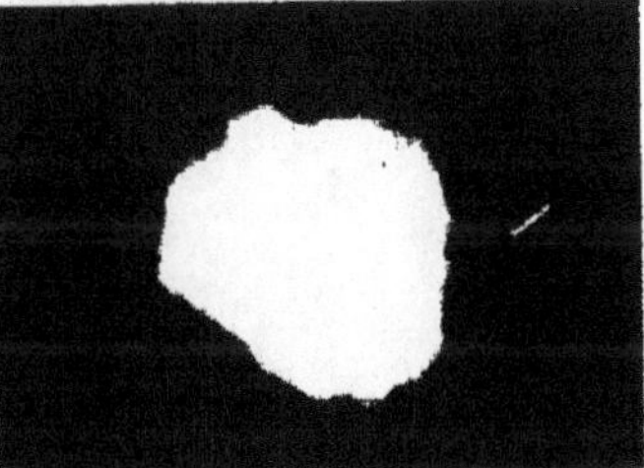

Russians say this (left) is Lunik and white splotch (above) is rocket's sodium flare. Author Mallan brands both Russian-released photographs as fakes.

The BIG RED LIE

sions, ordered merely another vaguely puzzled essay.

These men are on the trail of the Big Red Lie, but unfortunately they must piece the story together from shadows on this side of the Iron Curtain. There is virtually no more chance of getting any useful leads inside Russia, for the Kremlin is now undoubtedly on guard to protect its lie. I was lucky enough, though, to get the story *before* Lunik. I saw Russia and talked to Russian scientists at a time when there was no Lunik hoax to protect.

In my notes and pictures and tape recordings there is, I submit, solid evidence that no such sun-orbiting rocket—and no effective Russian ICBM—could or does exist.

I arrived in Russia one year ago this month. In my luggage were letters to top Soviet scientists from their counterparts in U.S. science—men such as physicist Dr. S. Fred Singer of the University of Maryland, the man who first seriously proposed an earth satellite and showed that the idea was feasible. These letters asked the Russians to co-operate with me, identified me as a trained scientific observer and author of scientific books, and stated my mission as that of bringing home an objective and, presumably, flattering report on the state of Soviet science.

The whole idea of this mission sat well with the Russians, for they are politically committed to brag. To a huge extent, the success of Communism depends on the success of Communist propaganda in winning the world's admiration. From Premier Khrushchev down to the lowliest newspaper hack, Russians lose no opportunity to tell the world of their every achievement. I fitted into this effort very nicely. Here I was: an American science journalist, all starry-eyed, eager to add my voice to the admiring chorus. I really was. As I entered Russia that spring, I was fully prepared to be overwhelmed by an impression of tremendous technological progress.

I let the Russians know this, and they welcomed me. I have no other explanation for the fact that I was allowed to talk with so many key scientists, to see so much of what previously had been obscured. American news correspondents whom I met in Moscow gaped in disbelief when I showed them the list of Russians I planned to interview. They told me that they rarely, if ever, get to see a Soviet scientist of any note. Most of their science news is spooned out to them by Soviet press bureaus. It isn't hard to guess why: these are hard-digging reporters, and the Kremlin fears them. I, on the other hand, was billed as a man so full of awe and admiration that I'd swallow whatever was handed me.

Intourist, the Soviet government travel bureau, assigned an intelligent, moderately pretty girl named Natasha as my interpreter and guide. I told her that the first man I wanted to see was Prof. Leonid Sedov, head of the Permanent Interdepartmental Commission on Interplanetary Communications (*i.e.*, space travel) of the Soviet Academy of Sciences. As over-all boss of the Russian space program, Sedov had two things that I needed badly: (1) the locations and phone numbers of other important scientists, and (2) the power to help me, or not help me, get in touch with them. There are no phone books available to the public in Russia, and it is very, very hard to track down key men without help. Hard? Impossible.

Sedov, a big man with thick rimless glasses, was as obliging as a salesman with a hot prospect. I asked if I could look in on key research projects. Of course I could. Could I interview people working in the space program? Cer-

GENERAL LESLIE GROVES, former chief of the Manhattan Project: "I have always been skeptical of the reports we receive on Russian progress in scientific engineering and will remain so until there is free access across the Iron Curtain."

Amateur photographer in Scotland took this picture of a light in the sky after Lunik "launching." London papers said bright area might be a flare, but experts disagree.

Small Russian radio telescopes such as this were said to have picked up signals "loud and clear" from Lunik, whereas . . .

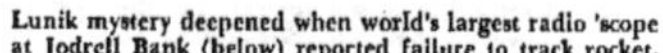

. . . giant U.S. Air Force instruments like this one (left) had no success in locating any trace of the fugitive Red rocket.

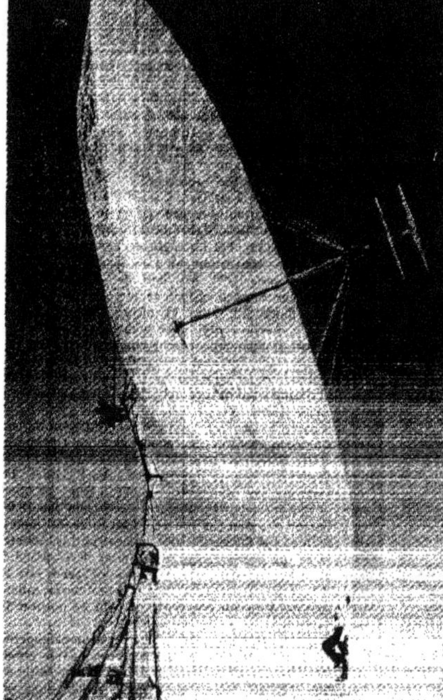

Lunik mystery deepened when world's largest radio 'scope at Jodrell Bank (below) reported failure to track rocket.

Author Mallan, right, and girl interpreter chat with astrophysicist Prof. Gleb Chebotarev (second from left) and aide.

The BIG RED LIE

tainly. Could I photograph rockets? Sure thing. I handed him a list of the top men I hoped to visit. Would he tell me where these people were and smooth the way for me to see them? He nodded amicably, jotting reminders to himself in a notebook. It could all be arranged. No problem.

Obviously, then, the Russians intended to show me things, send me reeling back to my hotel dazzled by scenes of mighty rockets and pioneering experiments. Mentally, I shrugged. It was an acceptable bargain. If they wanted to put on a show for me, I'd give it honest reviews when I got home.

But almost as soon as the curtain went up, I sensed that something was wrong.

One of my key interviews was with a man who, as far as I know, had never before talked with a Western reporter: Prof. Gleb Chebotarev, director of the Institute of Theoretical Astronomy in Leningrad. Chebotarev is one of Russia's top mathematical astronomers. In the months before I'd come to Russia, he had been widely touted in the Soviet press as the head of "Project Boomerang," a planned effort to send a rocket around the moon. According to these reports, Project Boomerang was well along; the Russian moon rocket would be launched quite soon.

Chebotarev was a bespectacled, amiable, middle-aged man; he looked the way astronomers look in the movies. I asked him how close his project was to its goal.

He shrugged. "I cannot say," he said. "I have not yet been approached by the engineers."

Trying not to look startled, I asked: "At any rate, your project is being seriously considered for future use?"

"Right now this is only theoretical work."

"You mean to say that no engineers have read your calculations and at least commented on them?"

Chebotarev smiled. "If they have, they have not let me know. I do not think they believe in my sanity."

It was a shock. All along, I'd pictured the Boomerang project as a missile center alive with busy men, designs being drawn, components arriving, maybe a launching pad being made ready. Instead, it turned out to be a middle-aged professor in an ivory tower. The propagandists had grabbed a little weed of fact, held a magnifying glass up to it and made it look like a tall tree.

I was badly shaken, but I tried to hide it. I went on with more questions. Before coming to talk with Chebotarev I'd read over some newspaper clippings on him, and I now remembered a fragment from the New York *Times*. The *Times* was quoting from a *Moscow News* story on Boomerang: "The increase of a rocket's speed from the 18,000 mph already achieved (by the Sputniks) to 25,000 mph so that it can escape the earth's gravitational pull is 'perfectly possible at the present stage of technical development.'"

I asked Chebotarev about the problem of speed. Said he: "It will be a very great jump from eight to eleven kilometers a second. I do not know when it will be possible for them to do it."

Another shock. I sat there and looked at Chebotarev, trying to figure it out. He wasn't lying to me. Why should he? When the whole Communist propaganda machine was bragging raucously about the coming moon shot, why

GENERAL NATHAN F. TWINING, Chairman of the Joint Chiefs of Staff, commenting on Russian Defense Minister Malinovsky's assertion that Soviet ballistic rockets can pinpoint their targets, has said flatly that he doesn't believe the Red claim.

GENERAL BERNARD SCHRIEVER, chief of the Air Force ballistic missile division, when asked whether he accepted Russian boasts about ICBM's, said: "I don't believe the Soviet Union now has an operational long-range missile ready."

should this one scientist suddenly try to play it down? If he was going to lie, he'd have lied in the other direction; he'd have told me how the rocket was poised and ready to shoot.

I could only conclude he was telling the truth as he knew it. He was a scientist. Scientists are rigorously trained against falling into the trap of exaggeration; they make statements cautiously. They know that, in their business, a man's professional reputation can collapse overnight if he's caught saying misleading things about his work.

Then I wondered: wouldn't Sedov have foreseen this? Sedov, or whoever else was masterminding my journey? Wouldn't he have realized that, in sending me to see scientists, he was sending me to see truth? Maybe not. Maybe he believed the propaganda himself, or maybe he felt certain that the scientists, being Russians, would loyally back up the Big Red Lie.

Or maybe, I thought, maybe there's a moon shot under wraps somewhere, and Chebotarev just doesn't know about it. But this didn't stand up. One of Russia's top astronomers, a well-known expert in celestial mechanics, kept in the dark about Soviet space activities?

I put out another probe. I remembered reading an article in *Znaniye-Sila* (translation: Knowledge is Power), a leading Soviet science magazine. The article dealt with Chebotarev's lab, and it said in part: "By means of huge automatic computing machines, this Institute is engaged in calculating the precise movements of the earth, moon, and planets . . ."

I said to Chebotarev: "Perhaps you can give me some information regarding your electronic computers?"

Chebotarev said: "I am not acquainted with these machines. In my kind of work the electronic machines are not needed. All my calculations are made by another professor with the help of small electrical calculators."

It was unbelievable. Not *Znaniye-Sila's* exaggeration, which was remarkable enough, but the fact itself. In the U.S. today, electronic computers are 100% essential both to theoretical astronomy and to the nuts-and-bolts work of space travel. Even small research shops use computers—scramble for them, rent them part-time when there's no other way to get them. Yet one of Russia's best astrophysicists is "not acquainted with these machines."

I wanted to read more about Soviet computers. I knew that a good place to do so was at Moscow's Technical Bookstore, the one place in the city where scientists, technical people and students can buy their specialized books. In this store, I reasoned, I'd find descriptions of the most up-to-date Soviet equipment.

I sneaked in without Natasha, for I didn't want to give her too many clues to the direction of my thinking. I have only a minimum knowledge of the Russian language, but mathematical and scientific symbols—and, of course, pictures—know no language barriers. In books with 1957 and 1958 publication dates, I found diagrams and photos of primitive electronic machines such as those used in America back in the early 1950's. In the fast-moving computer business, six or seven years add up to a long, long time. Compared to what IBM and Remington Rand are producing today, those old machines were [*Continued on page* 102]

EDITOR'S NOTE: Next month TRUE Correspondent Lloyd Mallan will reveal Other Big Lies which he discovered in Russia. Be sure to read his revelations on the myths of Soviet air power.

ANOTHER RED LIE...

UNIVAC

FIRST CHOICE OF INDUSTRY AND GOVERNMENT

Remington Rand Univac ad ran October 24, 1955.

***Red Star* used retouched drawing, August 23, 1956.**

ORDINATEUR SOVIETIQUE
(d'après l'Etoile Rouge — 23 Août 1956)

French publication later accepted Soviet fakery.

An interesting example of Russian techniques of deception is documented above. The drawing of a Remington Rand Univac computer (top) appeared as part of a full-page advertisement in the international edition of *Time*. Ten months later, the same drawing, now minus the brand names at the top of the computer, appeared in the Russian Army paper *Red Star* in conjunction with an article on Soviet progress in the computer field. As has so often happened before and since, the lie found some takers: a French technical journal, *Review of Automation*, soon picked up the *Red Star* story, faked picture and all, and passed it along as fact for the rest of the too-trusting world to believe.

48438 O—59——8

slow, not very versatile, and hounded by frequent breakdowns. I got the definite impression—and Chebotarev's statement backed me up—that even these machines were not used widely in Russia.

Maybe Russia is keeping its new computers under wraps, I thought. But this didn't make much sense. Keep them from Chebotarev? Keep them from the Technical Bookstore?

I was deeply puzzled. Without high-speed computers, you can't advance very fast in space technology. Maybe you can slam a satellite into a half-baked orbit by bolting a few big rockets together, aiming them at the sky, and hoping. But you can't quickly progress from there to the delicately instrumented, perfectly controlled craft that will land on the moon or pass close to it. You can't quickly progress to new, more powerful kinds of propulsion. Millions upon millions of calculations are necessary. Take nuclear propulsion. Both IBM and Remington Rand are now building for the U.S. Atomic Energy Commission machines that will multiply a pair of 15-digit numbers in millionths of a second. These machines will do as much work in a day as the fastest existing computers can do in a month—and even so, says AEC, don't expect results too soon.

Next, I went to Moscow's Polytechnic Museum, the most important scientific showplace in Russia. In a special room devoted to progress in electronics, I found a computer that U.S. scientists would not have been very proud of even back in 1952. This was the Polytechnic Museum, an institution designed for showing off. Could I assume this backward machine was an example of Russia's best?

I was reminded of the time a year or two back when a Russian magazine published a picture of a Univac, with Remington Rand's label blacked out and a caption proudly referring to Soviet progress in electronic brains.

The thing that impressed me most at the museum was something that wasn't there: miniaturization. In space flight, where every ounce of weight counts, you must build microscopically small. You must have tiny sensing devices, a tiny computer, tiny navigation and control equipment. Without such miniaturized innards, the U.S. Atlas missile, for instance, would not be able to guide itself through space and land on a target 6,000 miles away. That's controlled space flight, necessary either for an ICBM or a trip to the moon.

At the museum, where the pride of Soviet science is displayed, I saw no evidence of miniaturization. The smallest vacuum tubes on display were bulky compared to what's turned out in the U.S., Germany, England and Japan. The Russians are known to make transistors, tiny gadgets that can do the work of big vacuum tubes in certain uses, but none were on display. From this and from subsequent reading and looking, I got the impression that Russians don't consider miniaturization an important kind of progress.

They admire bigness. Premier Krushchev is fond of referring to the U.S. Vanguard satellite as the "Grapefruitnik." He alludes to the greater size of the Sputniks. What he doesn't say is that the Vanguard spaceball is evidence of a much more sophisticated technology, a much firmer mastery of miniaturization. The Vanguard was guided into an orbit so stable that it'll be up there for a good two centuries. Sputniks I and II had remarkably unstable orbits, indicating that they weren't guided but merely thrown almost haphazardly into space.

In fact, Dr. I. M. Levitt, director of the Fels Planetarium in Philadelphia, took the Sputniks' faulty orbits alone as evidence that the Kremlin has no ICBM, no really accurate guidance system.

The Sputniks are big. Could the reason be that Russian engineers don't know how to build electronic equipment small? U.S. scientists have grumbled that Russia hasn't published much data collected by the Sputniks. Could it be that there *is* no data—no fine enough instruments aboard to collect it? Even the instruments that are aboard seem inferior. The monster Sputnik III's signals have always been so weak that only the most sensitive receiving equipment can detect them. Tiny Vanguard, less than a thousandth as big, comes in loud and clear.

I came out of the museum with my head spinning. Over a quiet cup of bad Soviet coffee, I tried to reconcile what I had seen and heard with the picture of Russia that I'd carried into the country. The whole progress of world affairs in the past 10 years had been shaped, to a big extent, by the belief that Russia was developing ICBM's, fearsome weapons that could break up New York and San Francisco and Detroit in one colossal attack. Were these terrifying things nothing, after all, but figments of somebody's imagination?

No, I told myself. You're going too far on too little evidence. Surely, before long, they'll show you some instruments that'll knock your eye out.

I went to see Alla Massevich, Vice-President of the Astronomical Council of the Academy of Sciences. She heads Russia's optical tracking network. A Red newspaper had said that Sputniks and Lunik were tracked by "specially built scientific stations equipped with a great number of radio-technical and optical instruments." This is a field in which I have real interest and knowledge, and I was anxious to talk to this celebrated woman astrophysicist. Accurate tracking is a must; you need to know exactly how your spacecraft behave in flight in order to improve them. And for accurate tracking, you need extremely fine, super-fast cameras, radio telescopes, and other very precise equipment.

Alla Massevich took me up on the roof of the Sternberg Astronomical Institute. "We have twenty-five tracking stations like this at universities and observatories all across Russia," she told me, proudly.

The major piece of equipment was an ordinary Red Air Force aerial reconnaissance camera, adapted for its new space-track job. It was of World War II vintage. Sure, it was a good, serviceable camera—but if you tried to sell it at Cape Canaveral, they'd laugh at you. In fact, Edmund Scientific Company of Barrington, New Jersey, will sell you a war-surplus American aerial camera of the same type for under $80.

I asked Alla Massevich whether Russia's other tracking stations had the same equipment. She said they had. *All* of them? Yes. Later in my trip, I verified this by visits to the key optical teaching stations. I also talked to a scientist who was starting to develop more accurate space-tracking cameras. But his project had hardly got off the ground.

It was all very hard to believe. The whole thing had the air of a last-minute scramble to throw a tracking network together with whatever equipment happened to be lying around loose. In the U.S., scientists had planned ahead. The Navy had gone to Perkin-Elmer Corporation more than a year before the first satellites were due to be launched, and had ordered special tracking cameras built at a cost of some $100,000 each. These cameras weren't just set up wherever universities and observatories chanced to be, but were very carefully spotted across the Northern and Southern Hemispheres for the best possible tracking effectiveness.

Well, I thought, maybe the Russians depend on radio and radar more than optical tracking. I asked Dr. Massevich about this. Said she: "We have amateur radio enthusiasts all over Russia. They collect this information for us."

I pressed her further on the point, later asked other scientists about it. Crazy as it seemed, that's the way it was: this huge, science-minded nation depended on hams to keep track of the satellites' signals. The U.S. welcomes hams' tracking reports, but these aren't enough for the precise measurements demanded by an advanced space technology. In the American space program, official, professionally manned radio tracking posts are located at carefully picked spots on the map.

I remembered something about the Sputniks. Instead of broadcasting on the high frequency bands originally agreed on by International Geophysical Year scientists, the Russian moons came in on low frequencies—the frequencies allotted by worldwide convention to hams. This surprised and puzzled the West. Maybe, I now thought, the reason was simply that Russia didn't have enough of the more critical, harder-to-build high frequency equipment. Maybe Russia had to depend on ham receivers because nothing else was available.

I also remembered something else. Some three months after Sputnik I was launched, the Russians lost track of it. An official announcement from Moscow said that it had burned to nothing in the atmosphere. This revealed much about the precision of Russian tracking equipment. Eight days later, astronomers at Ohio State University's Radio Observatory were still tracking the fragments of Sputnik I that remained in orbit.

My whole concept of Russia was turning upside-down, and it turned a little more with each scientist I questioned. I went to see Prof. Alexander Mikhailov, director of the Pulkovo Observatory at Leningrad. A big front-page story in the

Approved For Release 2004/08/31 : CIA-RDP63T00245R000100290011-7

Moscow News had startled Western astronomers with this statement:

"The biggest telescope in the world, with a six-metre (236-inch) diameter reflector, is being built in the Soviet Union. The designing of such an astronomical instrument is a very complex technical problem . . . It took ten years to build a similar telescope with a five-metre reflector in the United States . . . The new instrument will help astronomers in their observations of artificial satellites and interplanetary rockets which may soon fly around the moon, to Venus and Mars."

That's a whopping big statement. In one blow, it (1) sneers at the present world's largest (200-inch) scope, at Mount Palomar, California, (2) establishes Russia as a nation with superb technical skills, and (3) hints at an ambitious space program. I was eager to hear more about this new giant telescope. I asked Mikhailov for details.

"That telescope will take ten or fifteen years to make," he said. "The construction has not yet begun."

I asked him for details on the design. There weren't any. "It is not yet decided what kind of mount it will have," Mikhailov said. "We do not know yet where it will be installed."

In other words, the telescope that was "being built" hadn't even been designed. About all that had happened, apparently, was that somebody had decided it would be nice to have such a telescope some day. Once again, the weed of fact and the magnifying glass.

This was how the Big Red Lie was built, I reflected. It was cumulative; each lie supported a complex of others. Having believed other Russian exaggerations about their technical progress, the West was ready to believe the giant telescope story. And, believing that, it now stood ready to swallow still more.

I tried to get an interview with Prof. Viktor Zhdanov, Russia's Deputy Commissioner of Public Health. He had been quoted in the Soviet press as a space physiologist. He was said to be studying problems of human survival as these might be posed by viruses in space during interplanetary flight. He turned me down. "I do not have connections in these fields," he said.

I had a long session with Professors Yury Pobedonostev and Kiril Stanyukovich, both key scientists in the Soviet space program. One topic I had on my mind was inertial navigation, an important technique used in guiding U.S. rockets and missiles. I was fairly sure by now that Russia had not mastered the technique, for it requires a miniature computer and other kinds of high-grade electronic engineering. But I asked any way.

I was right. "Some work in this field has been done," Stanyukovich told me. "But it is all theoretical so far."

I turned to another topic. The Soviet press had made much of Russia's intensive program to send men up in satellites and out to the moon. I asked about pressure suits, the kind that would be used by men in outer space.

Said Pobedonostev: "We will develop such suits when we need them."

Here it was again: the apparent unwillingness or inability of Russian space officials to plan ahead. In the U.S., pressure suits are the subject of elaborate and determined research, and have been for a good fifteen years. Both the Air Force and the Navy have already developed operational pressure suits.

I asked: "Are you working toward manned space flight?"

"Yes," said Pobedonostev.

"Then you must have rocket-powered aircraft?"

"We have jets."

"No—I mean pure rocket-powered aircraft such as the American Bell X-1 series."

"We do not believe in such aircraft."

I stared. The rocket plane, capable of climbing over the top layers of atmosphere, is an essential experimental tool for manned space flight. It trains pilots in the art of flying through a vacuum, gives space physiologists priceless information on medical problems involved. It's the forerunner of the true spaceship, the logical stepping stone to interplanetary travel. I pointed this out.

"Such flying," said Stanyukovich, "is much too risky."

I asked what Russia was doing about space medicine, in that case.

"We make use of dogs," Pobedonostev replied. "The dogs will lead to the time when we can use men. One cannot know where he is going until he gets there."

"But if you don't know where you're going until you get there," I pointed out, "then you must take risks. You claimed rocket aircraft are too risky, but aren't such risks necessary for achieving space flight?"

Stanyukovich suddenly exploded with anger. He and Pobedonostev held a heated argument in Russian, which Natasha didn't translate for me. Obviously, I'd touched a tender spot. Despite all the bragging, Soviet Russia was a long, long way from landing a man on the moon. Matter of fact, there seemed to be no organized, long range program at all for any such purpose.

Bit by bit, my picture of the Russian space and missile program was fitting together. The program, it turned out, was more noise than reality. It was a superb example of the propagandist's art.

I went on probing. I dropped in on *Znaniye-Sila*, the science magazine. Among the photographs that the editor proudly pulled out of his files for me was one of a radar-optical camera. It had a caption subtly leading the reader to believe the instrument was of Russian make. **I knew better. I'd helped take the photo myself four years before, at the White Sands Proving Ground in New Mexico.**

Academician Sedov, who had promised me the world when I started out, now seemed to have changed his mind. He'd disappeared, leaving me to shift for myself. I was getting nervous. More than once, I thought I was being spied upon—and whether it was just nerves or the real thing, it was enough to worry me seriously. I was also worried about my photographs and film, my tape recordings, my notes. I began to think of ways to get them out of there safely.

It was time to go home. The film and tapes were a real problem. By Soviet law, I was supposed to submit them to the government for review before taking them out of the country. But if I did so, I felt, I'd be unlikely ever to see them again.

I put the tapes, photos and film on the bottom of a large suitcase, surrounded them with socks, covered them with a layer of shirts. On top of the shirts I put a row of scientific and technical books that had been autographed for me by their authors—big men like Pobedonostev and Chebotarev.

I walked up to the baggage inspector at the Moscow Airport and handed him the hot suitcase first. Then I held my breath.

He lifted the lid. A frown crossed his face when he saw the books' titles—all severely technical. Brows knit, he picked up one of the books, flipped through it. He looked at me, then back at the book. He seemed to be wondering whether it was all right to allow such books out of the country. It was, of course, but now I'd made him suspicious and alert.

Finally he saw the flyleaf and the autograph. His eyes widened. He looked at the other books' flyleafs. His hand wandered over the layer of shirts. He was going to lift them up. I thought: I'm finished.

Then he pushed the suitcase at me. I'd made it.

In West Germany, I was questioned about what I'd seen by U.S. intelligence agents. Eventually, I got home. The American picture of Russia hadn't changed much since I'd left. There was still fright in the air. Russia was dangerous, they said. Russia claimed to have intercontinental missiles that could cripple America in one crushing blow. Russia would soon have armed satellites and a base on the moon. Russia would control space and all earth beneath.

This wasn't the Russia I knew.

The Russia I knew was not capable of designing or building an effective intercontinental missile—or, indeed, any long-range rocket that would know where it was going before it got there. This Russia was a good twenty years behind the U.S. in space technology. It had little more hope of controlling space than did the Principality of Monaco. If it was ever foolhardy enough to start a war, the Strategic Air Command would destroy it as a nation in one week.

Then the Soviet Union announced Lunik. You may imagine how it hit me. Only by an incredible freak of luck could a Russian rocket have turned in any such performance. A moon shot predicates the existence of highly sophisticated computers, clever miniaturization, precise tracking, fine guidance, high-grade electronics. These things were not available to the scientists who allegedly built Lunik. A moon shot must also be planned very carefully in advance. No such plan existed when I was in Russia half a year before. One lonely professor was mulling over the idea. No one else seemed actively interested.

Lunik had to be a hoax. I was almost certain of it. I started to check around.

By the time I finished, I was 100% certain.

The first new clue I got came, paradoxically, from a phone call that didn't go through. The Lunik announcement reached the U.S. on a Friday evening. On Saturday, I placed a call to Ohio State's Radio Observatory to find out whether any signals had actually been heard from Lunik. The switchboard operator told me that the astronomers had closed shop for the New Year's holiday weekend.

Sure, I thought. If I wanted to perpetrate a space hoax, when would I do it? Over a holiday weekend, when the Western World was relaxed.

On Monday, the Kremlin announced that Lunik's radio transmitters had gone dead as it passed the moon and headed into an orbit around the sun. Thus, all contact with the rocket–and all hope of proving or disproving its existence–was thenceforth and forever lost. Convenient. A little too convenient, perhaps. Two months later the successful U.S. space probe, Pioneer IV, kept transmitting useful data on three channels until it was almost half a million miles away.

When I finally got in touch with the Ohio State people and other tracking stations, a provocative new fact came to light: nobody, absolutely nobody, had received a signal that he could identify with certainty as coming from Lunik.

The Russians claimed that, on their receivers, signals were coming in clear as bells on all four of Lunik's announced radio frequencies. No one else had any luck, although trackers throughout the world had picked up Sputniks, U.S. satellites and moon probes–including, later, Pioneer IV. The huge radio telescope at Jodrell Bank, England, which later followed out beyond 400,000 miles, and which detects radio impulses from galaxies thousands of trillions of miles away, scanned the skies for 18 hours without finding a trace of Lunik. Prof. Bernard Lovell, Jodrell Bank's director, was quoted by the New York *Daily News* in March as speculating that "the whole story may have been a Russian propaganda hoax."

Other trackers around the world had doubts, too. Drs. Nakada and Takeuchi, top Japanese tracking scientists who had caught signals from all other space vehicles, detected nothing from Lunik. "We have to rely on the Russians' claims," they told *Astronautics*, official publication of the American Rocket Society. "We cannot acknowledge the existence of the rocket on the basis of our own findings."

According to George Grammer, official of the American Radio Relay League, the search for Lunik was "a complete negative as far as we're concerned." The ARRL has 90,000 members throughout the world, and hundreds of them reported picking up each U.S. and Russian satellite. But only one report came in on Lunik. The ARRL is now sure the signals described in this report came from Sputnik III.

I checked with military and intelligence men to find out what they thought. They were wary; I sensed that they wanted to keep their doubts to themselves. But the doubts were real. One Pentagon general told me: "Yes, it could have been a hoax. There's at least one other case where we know they were doing the same thing." He wouldn't elaborate. There was a good deal of embarrassment in the air. Another general smiled enigmatically. "Even if I knew," he said, "I wouldn't tell you." I guessed that he knew.

The embarrassment seemed to stem partly from the fact that President Eisenhower had publicly congratulated the Russians. He'd look silly, the feeling ran, if he now had to retract his words. But still the doubts piled up in secret. I was amused to notice that it took weeks for any of this to leak into the newspapers. On January 21, Fulton Lewis Jr. wrote in his column: "Allen Dulles, chief of the Central Intelligence Agency, reported ruefully to a secret session of Senators and Representatives that all evidence indicates the missile never went beyond the general vicinity of the moon, if, indeed, it got that far." Shortly afterward, the Defense Department slammed a security lid on all information about Lunik, and the papers lapsed into silence on the subject.

But the doubts and puzzlement continued. I talked with John T. Mengel, Director of the Tracking and Guidance Section of Project Vanguard. He told me that, theoretically, the skin temperature of a rocket or other body in sunlight should grow hotter as it moves away from the earth. Thus, the Lunik on its sun bathed way to the moon should have reported considerably higher temperatures than does the Vanguard satellite, which never goes farther from the earth than 2,460 miles. But the people who invented Lunik apparently didn't know about this. An early report from Tass, the Soviet news agency, boasted that Lunik was 130,000 miles from the earth and stated that its skin temperature was 59 to 68 degrees Fahrenheit. In months when the Vanguard is in sunlight all the way around its orbit, its skin temperature goes up near 150 degrees.

To bolster their lie, the Russians released what they said was a photograph of Lunik before launching. At least two things about this picture made it suspect. First, it was very crudely and very heavily retouched–as much a painting as a photograph. Second, the transmitting antenna system depicted could not possibly have broadcast clearly from outer space on Lunik's announced wavelengths. According to Prof. Thomas A. Benham of Haverford College, well known expert in radiophysics, the Russians would have needed a receiving antenna system several acres broad to catch any signals at all–let alone the loud, clear ones they claim to receive–from such a rig.

The Kremlin also published what it said was a photograph of a sodium flare released by Lunik at a point 62,000 miles out from earth. The photo, allegedly, was taken by the Alma Atá Observatory in Soviet Asia. It wasn't very convincing. It showed a great big nondescript white blob on a starless black background. There was no grain in the photo, so it wasn't an enlargment. I'd visited the Alma Atá Observatory, and I know that the biggest telescope there is 500 millimeters in diameter. Even if Lunik's flare was a mile across, it would have appeared to this telescope as a mere dot of light from 62,000 miles away.

Only one non-Russian observer thought he saw Lunik's flare. This was a Scottish cameraman, who snapped a picture of a cigar-shaped patch of light in the sky over Edinburgh. If it was 62,000 miles away, it would have had to be at least a thousand miles across to appear that big to the camera's naked eye. Scientists, examining the picture, concluded that what the Scotsman saw was really the result of a temperature inversion–a mirage-like reflection of light from the city below.

Yet despite all these inconsistencies and improbabilities, the U.S. man-in-the-street believed in Lunik–just as he has believed other parts of the Big Red Lie. Americans in general are gullible when it comes to Soviet stories; no boast from that big, mysterious nation seems too fantastic to be true. The Lie is self-supporting.

Occasionally, you hear a voice beseeching everybody to calm down. Late in January, for instance, General Nathan F. Twining, Chairman of the Joint Chiefs of Staff, stated flatly that the Kremlin's boast about having operational ICBM's is just that–a boast. Said Twining: "There's nothing to it."

But most Americans fear Russia. It is easy to fear something you don't understand; and to most Westerners, Russians are a strange, paradoxical people. They don't think or act like us, as many Western reporters have noted. John Gunther, author of *Inside Russia Today*, comments that the Russians can build big, powerful machines, "but a simple flashlight seems beyond them." It's hard for Americans to understand this. We tend to think that, because the Russians can blast an earth satellite into orbit, they must also be able to guide a rocket to the moon or a missile to a U.S. city. It just isn't so.

It is also hard for Americans to understand how anybody could have the sheer nerve to lie on so huge a scale. To the Russians, it comes easy. The lie, the hoax, the con game occurs again and again down through their history.

For instance, every schoolboy who has studied Russia knows of the famous hoax that was perpetrated on Catherine the Great in 1787. When she visited what is now the Ukraine in that year, the Governor General of the region, Grigori Potemkin, was worried over what might happen to him if Tsarina found out how poverty-stricken the area had become under his inefficient administration. To fool her, he had the wretched hovels along her river route daubed with bright new paint. Collapsed roofs were covered with painted cardboard. The hordes of ragged poor were ordered to stay indoors out of sight. The blighted landscape was even brightened with cardboard trees.

That's what Russians are like. Today, they're hoaxers not only by temperament, but also by political decree. Make no mistake, the Communists in control of Russia are without morality as we know it. They gibe at democratic nations for their honesty. As it was written by Lenin, the Father of the Russian Revolution:

"We repudiate all morality taken apart from human society and classes. We say that it is a deception, a fraud, a befogging of the minds of the workers and peasants in the interests of the landlords and capitalists.

"We say that our morality is entirely subordinated to the interests of the class struggle of the proletariat."

What Lenin meant, simply, was that if lying was useful to Communism, then it was moral. To those in the Kremlin today, lying is not only useful; it is a foundation stone of their power.

Next time the Kremlin announces a new space triumph, consider carefully whether the triumph is real–or whether the Russians are merely, once again, practicing the gospel that Lenin preached.–Lloyd Mallan

REPRINT FROM

TRUE

THE MAN'S MAGAZINE

MAY 1959

Mr. ANFUSO. Then I would suggest, Mr. Chairman, that Mr. Mallan give to this committee the scientific precautions which he took in asking questions, some of the data that you are familiar with, how you evaluated it, so that when we have other scientists, they can either say you are right, or you are wrong.

Mr. MALLAN. Fine. Shall I proceed?

The CHAIRMAN. Proceed.

Mr. MALLAN. Again as I mentioned previously, I am thoroughly familiar with our missile research and the development methods we use. In order to perfect a successful missile as distinct merely from a rocket engine which requires a lot of "sweat," to put it in Air Force terms, there are two basic requirements: One is an optical requirement and the other is an electronic requirement.

In order to know how the exhaust flame of a rocket or a missile operates in the various layers and pressures as it gets higher and higher in the atmosphere, you have to continually photograph its exhaust flame at the exact moment of timing and also in order to note the stability of the missile. At the same time, electronic equipment inside the missile is telemetering information back to receiving stations on the ground which analyzes the temperatures, the rate of climb, the stability, the rate of change of velocity—many complex things—and then these are compared with the optical features and photographs. Out of this—and although it is very complex and I am simplifying it—out of this you learn after that one rocket shoot how it could be improved, even if it is an apparent complete failure such as a rocket blowing up on the launching pad.

This is still noted electronically and optically, and the reason for the failure is then determined and the reason is eliminated. This is how you develop a missile.

On the other hand, to develop a spaceship you first have to develop missiles; that is a preliminary step—rocket engines and missiles. A missile can be a guided rocket, or a guided turbojet, or a guided anything, as long as it is something that can propel and guide itself to a target. Also a spaceship has to take itself to a target and it requires at least a rocket engine.

You cannot use an air-breathing engine on that because it is operating in a near vacuum. You also have a human element involved, the pilot.

In the case of our rocket-powered aircraft, again electronics and optical equipment are an essential part of the development of the program.

The X-15 rocket-powered aircraft which soon will be running powered flight tests could never have existed without the earliest test of the XS-1 back in 1947, when Capt. Chuck Yeager first flew through the sound barrier in level flight after some of the foremost aerodynamic scientists in the world said this was impossible. It was made possible because the U.S. Air Force was able, with its technical background and know-how in physiology—they had just developed a partial pressure high altitude suit and helmet which Yeager was testing at the time that he was testing the airplane—to produce the first pure rocket-powered aircraft in the world, as far as I know

From this step, which was preceded by 10 years of planning and research, came the X-2 and——

Mr. ANFUSO. Excuse me, and Mr. Chairman, up to that time, up to 1947, there had been no evidence that the Russians had made that progress.

Mr. MALLAN. None whatsoever as far as I know. According to the Soviet scientists I spoke with, they as yet have no rocket-powered aircraft.

Mr. ANFUSO. Go ahead.

Mr. MALLAN. Well, I will try to be brief.

The point I am trying to make is that out of the background in physiological and biophysical research in altitude chambers, on centrifuges where very dedicated American scientists were often risking their own necks in order to find out things about unknown quantities in high velocity flight, with this background that had extended for about 10 years earlier before Yeager broke through the sound barrier, out of the background, I mean, came Yeager's flight.

Out of Yeager's flight came the flight of the X–2 rocket-powered aircraft which, even as Yeager was flying through the sound barrier, was on the drawing boards and from what he learned on his flight—that epic flight—the X–2 was modified and was designed to pierce what is called by some the thermal barrier, by some the heat barrier, and by scientists the thermal thicket.

This is the point of velocity of an object moving through the earth's atmosphere where its shape compresses the molecules of air to such a density that it is like boring through a brick wall. This creates friction which, in the case of aircraft or spaceships, would heat the ships up.

So the X–2 penetrated the thermal thicket and out of this came more information. The high speed flight test station—it was then called NACA, not NASA—out at Edwards Air Force Base was continuing quietly with no fanfare whatever to investigate the thermal thicket with the X–1–E which was a modified XS–1.

It is interesting to note that I saw in a Russian book a pen-and-ink drawing of the D–5582, the Navy's aircraft, the first aircraft that ever flew at twice the speed of sound. The caption said: "Rocket-Powered Aircraft Like These Are Being Used To Explore the Frontiers of Space." There was no credit line to the U.S. Navy, no credit line to the United States.

I found this kind of thing being done all of the time in Soviet publications. Then, digressing at this point——

Mr. ANFUSO. Excuse me, Mr. Chairman.

In other words, anyone in Russia reading that magazine or that article which you got from the library could get the impression that the Russians had invented it?

Mr. MALLAN. It seems to me that that impression was planned. Because I then began noticing it all over the place, in newspapers, magazines, and other books.

For example, I think 10 years ago Collier's magazine had a drawing of a proposed three-stage spaceship designed by Dr. Wernher von Braun. This exact spaceship, redrawn in pen and ink, appeared in two books, I have them at home and I had the captions translated.

They said merely that: "This is the three-stage spaceship which one day will take men to the moon." This occurs all the time in the Soviet press and in Soviet magazines and in Soviet books.

These are part of the things which led me to suspect the whole setup. At any rate, I was trying to give a general idea of how much effort, cooperation, coordination, scientific knowledge, and pure scientific research as well as courage are required in order to develop one rocketship or one spaceship, or one phase, like breaking the sound barrier, investigating the thermal thicket. Now the X–15 is ready to investigate regions 100 miles above the earth at velocities in excess, I am sure, of 3,000 or 4,000 miles an hour.

The point I am trying to make is that all of this did require and is still requiring miniaturized electronic components. Even in the early rocketship that Chuck Yeager flew through the sound barrier, there was a camera focused over his shoulder that automatically took four frames per second of his instrument panel. There were sensing devices on the elevators and sensing devices on the other movable controls of the aircraft.

There were cameras and electronic devices all over that airplane and they are all over the X–15 and all through it. These are the things that supply information to the ground stations that give the scientific background required.

In the Soviet Union I never saw a single miniaturized electronic component and what they did call miniaturized was clumsy, very heavyhanded.

They were using transistors, they claimed, in some of the sputnik instruments, but they put these instruments on display and the instruments—they might as well have used vacuum tubes in them, because they were very clumsy and large.

This does not make for efficiency and efficiency is exactly what is required in order to carry forth successfully a space flight program.

It is also required to develop an intercontinental ballistic missile. It is required to develop any kind of missile that can travel beyond a couple of hundred miles; I mean, beyond what the V–2 did or what an advanced type of V–2 can do.

I saw no optical or electronic equipment in the Soviet Union that could remotely compare with what we have here. In fact 25 years ago I was a radio ham myself and I had equipment the equal of the equipment I saw at the astrophysical observatories in the Soviet Union.

As another example, you can go to any radio shop in New York City and buy a ham receiver which would be far superior to a receiver I saw at the Russian Observatory.

They were proud of this. It was a remade receiver. They told me they got it from the Red navy and they let me listen to it. They were very pleased they could pull in the British Broadcasting Co. and let me hear some English on this. These scientists were very straightforward, nice people. I do not want to give the wrong impression about them.

Mr. Anfuso. Mr. Chairman.

The Chairman. Mr. Anfuso.

Mr. Anfuso. Mr. Chairman, are scientists generally—whether they are Russians or Americans or any other nationality—known to tell the truth?

Mr. Mallan. That is the basic tenet of all scientists. In fact, they hesitate to exaggerate.

Mr. ANFUSO. Is it also not true that the Soviet Union announced in January it would soon release all information relating to and collected by lunik?

Mr. MALLAN. Yes, they did announce that.

Mr. ANFUSO. Did they, after making that announcement, ever release the information?

Mr. MALLAN. No, they did not.

Mr. ANFUSO. You are positive of that?

Mr. MALLAN. I am positive of only one thing they announced since then. When our first Discoverer satellite was put into orbit—a very difficult orbit to achieve—about 3 or 4 days later—and, incidentally, I did expect them to come up with something to contradict this or to show that they were ahead of us—they announced that they had gathered some new information on cosmic radiation but they did not say what it was and the most significant thing——

Mr. MITCHELL. Will you yield?

Mr. ANFUSO. I yield.

Mr. MITCHELL. Mr. Mallan, did they not release some data?

Mr. MALLAN. No data. No data whatever.

Mr. MITCHELL. What did they release?

Mr. MALLAN. A general statement made at the Moscow Planetarium.

Mr. MITCHELL. No data was included.

Mr. MALLAN. No scientific data. Of this I am certain.

Mr. ANFUSO. Again, to complete the record, this information which my colleague has just mentioned, released in April; is that very extensive?

Mr. MALLAN. You mean the Pravda report of the lunik?

Mr. ANFUSO. Is that what you had reference to, Mr. Mitchell, the Pravda report?

That was issued in April?

Mr. MALLAN. I forget the date. No, it was issued prior to April.

Mr. ANFUSO. Well, this release which you say they made with reference to the lunik after January, I do not know exactly what time; is that available to you?

Mr. MALLAN. I believe I have it at home in my file.

Mr. ANFUSO. Can you submit that for the record?

Mr. MALLAN. I certainly am quite willing to.

(The information requested is in Russian and will be found in committee files.)

Mr. TEAGUE. Mr. Chairman.

The CHAIRMAN. Mr. Teague.

Mr. TEAGUE. Mr. Mallan, was it May Day a year ago when a whole family of Russian missiles was shown?

Mr. MALLAN. That, Congressman, was on November 7, the anniversary of the—the 40th anniversary of the Russian Revolution.

Mr. TEAGUE. November 7, a year ago.

Mr. MALLAN. 1957.

Mr. TEAGUE. At that time I questioned a number of men who were considered our top men in the Pentagon and asked, from pictures, Can you tell whether those missiles are missiles or whether they are something else? They told me in no uncertain terms they were missiles and they could tell that from certain things. Do you believe that to be true?

Mr. MALLAN. I am not quite sure that I believe that to be true. I believe that some of those on display were missiles in the sense that they were short-range rockets guided like the old-type German V-2's and which the Russians acquired over a thousand, including the plant where they were manufactured.

But to me most of those on display, and the largest one particularly, had a configuration like the Viking research rocket and the Aerobee research rocket of the United States, the data for which can easily be obtained, with engines that appeared to me from the photographs to be oversized V-2 engines.

Mr. TEAGUE. They would not make those big tracking vehicles tracking those missiles, that is one of the points our people made. Here they have this big family of missiles and they are not pulled by trailers as we are doing it but they are on a big tracked vehicle.

Mr. MALLAN. I do not see why they would not go to any amount of expense to build a prototype or two to show the world that they have something and then they boast that they have many of these things.

A perfect example is the delta wing aircraft that NATO has designated "Fishpot." When Gen. Nathan Twining, General White, and General Byrd were invited to the Soviet Union in June 1956 to witness Aviation Day, which is an annual event in the Soviet Union, out of the blue came these three delta-winged aircraft. They made one pass and disappeared.

I am sure, if you asked General Byrd—he is now retired, but he is a real expert on aircraft and he lives in Washington—his impression of these, in fact, if you ask General Twining, I am sure they will both say that these were very unstable aircraft.

Mr. TEAGUE. What about the jet transport that flew into England? Don't you think we were completely shocked and surprised at that?

Mr. MALLAN. The TU-104, yes; we were shocked and surprised, and I later flew in that same aircraft and it is not all that it is cracked up to be. As a matter of fact, it is a quick conversion of a twin-engine bomber which originally was a combination of components from the B-29-A-4-F which crashlanded at Vladivostok during World War II and which the Russians never returned to the United States.

Mr. MILLER. Do you know that?

Mr. MALLAN. I know that as a fact.

Mr. MILLER. That the Russian jetplane is made out of parts copied from the——

Mr. MALLAN. Not parts but it is a combination of design components that come from the B-29-A, the instruments are almost identical, the flight deck, you may check the Boeing chief engineer on this because when he saw the flight deck of the TU-104 that he felt very much at home on that.

Mr. ANFUSO. Is our jet much better?

Mr. MALLAN. Our jets are much superior. Our entire aviation industry is superior from what I know of aviation. I have a set of photographs here comparing the Electra, the Lockheed Electra, with one of the newer, in fact, the newest I think, Russian prop jet.

The CHAIRMAN. Do you have those photographs with you?

Mr. MALLAN. Yes; I do.

The CHAIRMAN. Do you want to pass them around and let the committee look at them?

Mr. MALLAN. I certainly shall be glad to.

Mr. ANFUSO. Could we have them made part of the record?

The CHAIRMAN. Mr. Fulton.

Mr. FULTON. May I make a suggestion that we have the witness make his statement on the facts that he has and the conclusion he has drawn to show that the missile flight announced on January 2, 1959, by the Russians was a hoax?

I would hope that we would get that in the record first so that we get his views before we get off into the development of planes and progress generally.

My point is, let us check the article on what his statements are, ask him about them; let him give his reasons for having made the statements and also his conclusions and the scientific evidence to back him up.

If I may with courtesy do it I would suggest that that be the procedure from here on.

The CHAIRMAN. Well, we have the witness who is the prime witness in this matter before us. I think it is wisdom on our part to get all of the facts as quickly as we can from him.

Mr. MILLER. Mr. Chairman, could we confine those to the statement about lunik; that is all. We have been getting out into foreign fields about which there is a great deal of controversy.

I think there are other people as prominent as the witness who have made statements as to the relative value of Russian planes. Now, he has made the statement about lunik. Let us keep on lunik, and see if he can justify his position.

Mr. OSMERS. Mr. Chairman.

The CHAIRMAN. Mr. Osmers.

Mr. OSMERS. Mr. Chairman—I read the article written by the witness and I must say that it is a very fascinating theory which he presents.

Unfortunately, Mr. Chairman, I am drawn to the same conclusion as the gentleman from Iowa, Mr. Wolf, that it would be entirely possible for an American science writer to go to Russia and not see one single thing that the Russians did not want to show him.

For example, on page 102 of the article, the witness, Mr. Mallan, describes a visit which he made to a politechnic museum. Now, I just cannot believe that it would be possible for an American science writer to walk into a Russian museum and obtain secret information any more than it would be possible for a Russian science writer to come and walk into the Pentagon or walk into some of our scientific installations and find access to top secret material.

Now, I would like to ask the witness this question: Even if we were to accept the statements which you have made as to their lack of development in miniaturization and tracking and telemetering, haven't they demonstrated an ability in rocketry to get a moon probe up into the vicinity of the moon?

Mr. MALLAN. With all due respect, Congressman, no; they have not demonstrated that.

Mr. OSMERS. Well, now, let us take this large sputnik. It weighed, I believe, in excess of half a ton; is that correct?

Mr. MALLAN. That is the weight that the Soviets announced or this officially.

Mr. OSMERS. Well, I have not heard it seriously questioned here as to the weight by any member of the scientific community and certainly I think we all recognize that no one can control what scientists say about such things. This committee has not heard it seriously questioned.

Do you question the weight of that sputnik?

Mr. MALLAN. Yes; I do.

Mr. OSMERS. Mr. Chairman, I still feel, as far as this committee is concerned, that this is an intriguing and an interesting article that is not, in my personal opinion, backed up by the type of scientific evidence that would be necessary to lead this committee to a conclusion. That is one reason why I would like to see the statements and the conjectures that are made here passed upon by scientific witnesses, the very best that we have. I do not think that we should brush it off, but there are some generalizations here which would have a lulling effect upon the American people that I do not think is warranted by other scientific Soviet affairs.

Mr. MALLAN. I would like to explain a very important thing, if I may, Congressman, and that is that a magazine has certain space restrictions and a certain audience.

I could not supply the complex technical data that would be necessary to answer the questions you want answered. However, I will be quite happy to answer them for you right now.

I would like to start by some quotes from prominent U.S. scientists.

Mr. OSMERS. Let me ask you this question: Recognizing the limitations that one American traveling with one Soviet interpreter has, in going through the Soviet Union, what leads you to believe that this great secretive nation would permit you, as an American science writer walking through under the guidance of one of their own, to see one single thing that they did not want you to see?

Mr. MALLAN. Well, it is very flattering to think that they would not want me to see anything. Also, simply because they are a secretive nation one scientist does not know what the other scientist knows and I cross-checked all my questions against another scientist—there were 38 altogether.

In addition to that, it would seem rather unlikely to me that they would go to the great trouble and expense just for me to move all of their fine equipment out of key positions and replace them with inferior equipment.

It takes a lot of sweat to move a radio telescope, for example.

Mr. OSMERS. You would not think, would you—I would and you apparently would not—that they would have an entire complex that was secret to the outside entirely and that they would have another complex that would be operated or maintained, shall we say, for visitors?

It would seem to me that that is what they would have.

Mr. MALLAN. I was not really a visitor. Of course, I am sure they have a secret complex.

On the other hand, there are certain ways that you can determine the technological status in these special fields from observing equipment at key installations.

They were not leading me. I requested, and then fought to visit these places. When I reached there in many cases the scientists there did not know I was going to be there until a day or so earlier.

Another thing is that I spoke with them on their own terms. Still another important thing is I was not accepted just as a reporter.

Most reporters—in fact, all reporters that I spoke with in Moscow from the free world, the correspondents there—were amazed that I could get to talk with 1 leading scientist let alone 24 leading scientists and 14 other scientists.

Mr. OSMERS. Mr. Chairman, that is one of the reasons that leads me to believe our Soviet friends may have been giving you what is known in the United States as a snow job.

On the other hand, I want to ask one more question of the witness.

The witness is obviously a man of the very highest intentions and has come here to offer to the Government whatever information he has. Now I want to ask him this question:

In view of what you learned in the Soviet Union and in view of what you know this country has done and is doing; what would you advise the Government of the United States to do that it is not now doing or would you advise this Government to stop doing what it is now doing?

Mr. MALLAN. That is a complex question in terms of an answer.

Mr. OSMERS. That is what makes being a Congressman difficult.

Mr. ANFUSO. Would the gentleman yield?

Mr. OSMERS. I would like to hear the question answered.

Mr. MALLAN. However, I do not advise that we stop in our defense effort.

For one thing, I discovered in the Soviet Union that the only thing that keeps the Government of the Soviet Union in place is the Strategic Air Command of the U.S. Air Force, and also I think that we should intensify our basic research program. I do not think we should cut our budget, I think we should intensify, reemphasize the program because it is the backlog of knowledge that counts, and also if we lose scientific leadership, we lose everything.

The Soviets have been able to gain a lot of leadership on the basis of propaganda. However, there was one other thing that I wanted to say and that is that I think we should take a firm stand with them. I think that we are allowing them to intimidate the world's foremost nation on the basis of sword rattling, on the basis of boasts, and on the basis, yes, of lies, really. Perhaps our people in intelligence know the true status of Soviet power, but on the surface our actions as a nation make us seem to be inferior to the Soviet Union as far as the rest of the world is concerned. There is this fear phobia which, although it may not be evident now, is having an effect on children.

Psychologically this is very bad for the kids of the United States of America. I do not think that we should retreat in our efforts. I think, as I repeat, they should be reemphasized. Science in basic research should be given more impetus and we should maintain the Strategic Air Command.

Mr. OSMERS. Narrowing that right down to the defense effort, in your opinion, Mr. Mallan, is this country spending too much for defense, not enough for defense, or about right?

Mr. MALLAN. Well, I do not really know. I really do not.

Mr. OSMERS. That sounds like an honest answer.

Mr. ANFUSO. Mr. Chairman.

The CHAIRMAN. I want to ask the witness two or three questions, if I may.

What equipment known to you would be capable of detecting signals of the strength, frequency, and other pertinent characteristics of those supposed to have been transmitted by lunik?

Mr. MALLAN. I think the equipment at Jodrell Bank, the radio telescope there, the one at Goldstone, the one at Stanford, the whole range of them controlled by the Air Force at its various missile ranges.

The CHAIRMAN. Well, we have a telegram from Stanford. Do you think that equipment is capable of doing it?

Mr. MALLAN. I think it is capable of it, but I do not think they did it. I think they heard something else.

The CHAIRMAN. Then you think Jodrell Bank in England is capable of the same thing?

Mr. MALLAN. That is the best equipment in the world for it right there.

The CHAIRMAN. Of the installations which you described there which did, in fact, they say, receive putative lunik signals on or about January 2 and 3 of 1959, did any of them say they received the signals?

Mr. MALLAN. Was this Stanford Research Institute?

The CHAIRMAN. Yes; and Jodrell Bank.

Mr. MALLAN. Jodrell Bank received no signals whatever and tried——

The CHAIRMAN. You think Stanford received signals but——

Mr. MALLAN. No, I do not think they were signals from lunik.

In fact, at this point, I would like to read into the record—and I do not want to be controversial but this was the missile and space activities joint hearing last January 29 and 30 in which this committee participated at which Dr. Wernher von Braun gave three basic proofs for the existence of lunik and it was published in the record.

I would like to contradict these if I may read a statement.

The CHAIRMAN. Before you do that, may I ask you a few more questions on the point?

Mr. MALLAN. Yes, sir.

The CHAIRMAN. Would you say that the failure of other stations to detect such signals was normal or unusual?

Mr. MALLAN. The failure of other stations to detect signals from lunik was absolutely unusual.

These consist of almost every tracking station in the free world. I do not really know how many, but there are at least 50 of them.

The CHAIRMAN. Was there any advance warning given of the lunik launching time, frequencies, and so forth?

Mr. MALLAN. There was about a 5-day advance warning.

The CHAIRMAN. So these stations had it.

Now, were the stations that received putative lunik signals any more sensitive, better situated geographically, or why did they receive it?

Mr. MALLAN. I can read you a quote directly from the director of the Goldstone tracking station—no, I am sorry, the signal expert at the Goldstone tracking station as to his reasons why they received them.

His name is William Tilkington.

The CHAIRMAN. Tell us what he said.

Mr. MALLAN. I will give you his exact quote. I asked him to explain how their station could have received signals from lunik when almost every other tracking station in the free world was unsuccessful and here was his answer:

> Our only explanation is we tried very hard and our sensitivity is very near the limit of sensitivity and we are in a very quiet area.

Then I asked him whether he was absolutely certain that the signals they received were from lunik, and his answer to that was:

> There is no way of being absolutely certain. We are fairly sure.

The CHAIRMAN. Now, let me ask you, Did Moscow indicate in advance what the frequencies would be?

Mr. MALLAN. Moscow announced four frequencies and the advance warning given to the free world tracking stations announced a range of five.

The CHAIRMAN. Now, the signals on return, were they the ones that had been announced by Moscow?

Mr. MALLAN. Do you mean in terms of the signals characteristic?

The CHAIRMAN. Yes.

Mr. MALLAN. Well, again Mr. Tilkington, who is the expert on signal characteristics at the Jet Propulsion Laboratory, when I asked him that question, answered that the pulse rate of signals was very different from that broadcast by radio Moscow as a tape recording.

However, there are or were similar characteristics to the signal with the one broadcast.

The CHAIRMAN. How strong were those signals?

Mr. MALLAN. Walter Larkin, the director of the Goldstone tracking station, when I asked him that question, answered it this way:

> From median width to nonexistent.

That is a direct quote.

The CHAIRMAN. Well, what duration did they have?

Mr. MALLAN. According to Mr. Larkin, over a period of about 5 hours they heard several bursts of signals, the longest of which was 3 minutes. But then Dr. Dryden, the Deputy Administrator of NASA sent me a graph plotted by Dr. Pickering, head of the Jet Propulsion Laboratory.

According to this graph there were four bursts of signals recorded in 5 hours and the duration of the longest one was about 10 minutes.

The CHAIRMAN. Well, did they seem to be Russian signals?

Mr. MALLAN. Dr. Pickering and Dr. Dryden both have made this flat statement, that there could be no doubt that these signals came from the lunik.

The CHAIRMAN. What data was obtained as to the bearings, bearing changes, the rate of change of the bearings and the angle between the apparent signal source and the moon, and the changes and rate of change of this angle?

Mr. MALLAN. Well, I can answer that quite simply. All of the changes were dependent upon an object in space that radiated signals.

I checked back in the "Nautical Almanac" for the latitude and longitude of Barstow, Calif., where Goldstone tracking station is located and discovered that at 2:30 a.m. on the morning of January 4, the moon rose and a little before 3:30 that same morning the planet Jupiter rose to the west of the moon and about 7° south of it, on the celestial sphere. This is where I think I should read my statement because I believe that all of the signal characteristics demonstrated on this chart are the characteristics of radio emissions from the planet Jupiter which, of course, is out in space beyond the moon.

The CHAIRMAN. Well, I want to ask you one more question and then I am going to quit.

On January 3, President Eisenhower sent a message of congratulations to the Soviets on this moonshot. Did he have independent evidence or do you know any existed at that time to confirm the Soviet announcement?

Mr. MALLAN. From what I know now there was no evidence at that time.

The CHAIRMAN. Any further questions?

Mr. Anfuso.

Mr. ANFUSO. Well, I am interested to know, Mr. Mallan, whether you think you were given a "snow job" or the American taxpayer was given a "snow job."

Mr. MALLAN. If it was a "snow job" it was the most expensive "snow job" the Soviet Government ever proposed or ever achieved rather.

The "snow job" would have required moving equipment back and forth across 14,000 miles. It would have been an exact contradiction to the party line of the Kremlin which is "We are a great big powerful nation and you better watch out for us."

Khrushchev is always talking about how advanced they are. So, it would seem unlikely to me that they would then show me equipment inferior to the equipment in the United States and another important factor is that the scientists in the Soviet Union have only to submit a need for new equipment. If they can submit new designs, there is no problem with the budget; they can get any amount for new equipment. Still this was the equipment you saw and far inferior to ours.

Mr. ANFUSO. And certainly the drawings that you saw in this library where they took credit for some of the things we had done, that certainly would not be part of the "snow job."

Mr. MALLAN. No; this was published before I got there.

Mr. ANFUSO. Mr. Chairman, I am not at all in disagreement with the gentleman from New Jersey. There is a strong probability that the Russians would not show this witness everything that they should show him and they certainly did keep an awful lot of information away from him.

I am inclined to go that far but I do believe this witness has made a very exhaustive study and we cannot treat lightly what he is saying to this committee.

I think that a very exhaustive investigation should be made. Would you be willing, Mr. Mallan, to submit to us a series of questions which you asked these Russians, these Russian scientists in order that we can ask those same questions of our American scientists and see how the answers compare?

Mr. MALLAN. I certainly would be happy to do that.

Mr. ANFUSO. I also wish to call to the attention of the committee that Mr. Mallan, when he did call upon me together with Mr. Daigh in New York, was not looking for any publicity as far as this whole investigation is concerned.

He volunteered to testify in secret sessions; is that not correct?

Mr. MALLAN. That is correct.

Mr. ANFUSO. To face any scientist who disagreed with him.

Mr. MALLAN. That is correct.

Mr. SISK. Would the gentleman yield?

Mr. ANFUSO. I yield.

Mr. SISK. Of course, the crux of the matter here, Mr. Mallan, as I see it, is that I do not understand how you can state with any certainty that you saw the key areas in Russia.

In view of a vast amount of testimony before this committee as well as other committees of the Congress by people who have visited Russia and spent time there, I am frankly at a great loss as to what you possessed either in the way of influence or anything else, that would have made it possible for you to have, in the first place, gone into their key areas and, in the second place, to know whether you were in their key areas.

I do not know to what extent you are familiar with what intelligence knows about the areas and some of the things that in closed session have been given to this committee.

Certainly to me that is the crux of the matter. I think, frankly—and I say this completely with all due respect to you and your knowledge—that you were really given a first-class "snow job."

Mr. MALLAN. Here comes that "snow job" again. I do not claim to have visited all of the key areas. I visited key areas in certain fields but there are fields vital to astronautics, to the field of space flight. Astrophysical observatories, for example, in this country are an integral part of our entire satellite program. The Smithsonian Astrophysical Observatory, for instance, is not only an optical center but it is a computing center and I visited the Crimean Astrophysical Observatory which is the major astrophysical observatory in the entire Soviet Union where they do their major research into the upper atmosphere as well as in earth magnetism, as well as in solar physics.

And electronic equipment is exceptionally important in these fields.

Mr. SISK. Well, I fully realize that, as I am sure the committee does. But the statement you are making, Mr. Mallan, is impeaching the testimony and the statements of some of the finest scientists in this Nation and the leaders of our missile and space program who have appeared before this committee time after time. You can understand the reason for some very strong and grave doubts that I might have about your statement when it does impeach the statements of our leading scientists.

The CHAIRMAN. Will the gentleman yield?

Mr. SISK. Yes.

The CHAIRMAN. I would just say, too, that I was reading one of the statements in this article to the effect that Russia does not have a long-range missile and it is a very emphatic statement.

Here it is here:

> Russia is not capable of designing or building an effective intercontinental missile, or indeed, any long-range rocket that would know where it was going before it got there.

Now, that sort of statement, if true, would have the effect of changing our entire defensive posture and, of course, I do not go along with that. My thought is that we should turn this matter over to a sub-

48438—59——4

committee and then proceed with the entire program of the progress made by Russia in science generally, military science, and other science. Also we have Prof. Thomas A. Benham, who had difficulty getting here this morning. I do not want to cut anybody off, but since we turn this over to a subcommittee, let Mr. Anfuso head the subcommittee, and we, perhaps, could ask Mr. Mallan to stand aside for a moment while we hear from Prof. Thomas A. Benham.

Mr. MILLER. Mr. Chairman, I have been comparatively silent today—that does not show dissent.

I just want to soften Mr. Mallan a bit by telling him that the Russians have done "snow jobs" on people that I consider are a little more astute than you are, in spite of the fact that you are a self-made scientist; so do not feel too badly about it, please.

Mr. MALLAN. But you are assuming that a "snow job" has been done and I cannot agree with you on that.

Mr. MILLER. Well, I do not agree with you, so we are in agreement on that.

Mr. TEAGUE. Mr. Chairman.

The CHAIRMAN. Mr. Teague?

Mr. TEAGUE. Mr. Chairman, I want to personally express my appreciation to Mr. Mallan for coming down here. I do not agree with my colleague, Mr. Sisk, that you have impeached our scientists.

I think you have disagreed with them. I have gone to a number of these so-called secret sessions and I have learned very little in them. I have come out shaking my head. I went to Russia and I came out of there so disturbed about our Intelligence, I made a definite effort to try to find out what we knew and I still do not know, so, I appreciate your coming down here.

The CHAIRMAN. On behalf of the committee, I want to say that the entire committee appreciates Mr. Mallan's attitude in coming down here and giving us the information that he has available. We appreciate it very much.

Mr. MITCHELL. Mr. Chairman.

The CHAIRMAN. Mr. Mitchell.

Mr. MITCHELL. May I make a brief comment?

I think from what has been pointed out here, your testimony is so inconsistent with the best Intelligence statements that this committee has heard in secret executive special sessions that it points out this to me.

You, as a lay person, as a reporter, can go into the Soviet Union and bring out some very strong opinions that we are so advanced as far as ICBM's, rocketry, and missiles are concerned, in general. It seems so inconsistent that you could have that firm opinion when our professionals get an entirely different opinion.

If you are right, what we should do is send a few roving reporters into the Soviet Union and do away with CIA—and thereby balance the budget. But I do appreciate your appearing, as do all of us on the committee.

The CHAIRMAN. We want to thank you very much, Mr. Mallan, and I would like at this time to call Professor Benham. Professor, if you would come forward?

Mr. DAIGH. Mr. Chairman, could I ask a question? The question here for examination is whether lunik exists or not, apparently; and, as Mr. Mallan mentioned in the record of your hearings, Mr. von Braun was asked a question. He gave three reasons why he thinks it exists. I think it is very pertinent at some time Mr. Mallan be allowed to answer Von Braun specifically on these three points that were raised.

The CHAIRMAN. I think this: We had better hear from Dr. Benham here and then we can go into that. I think that the subcommittee will give him every opportunity.

The CHAIRMAN. Professor Benham, will you hold up your right hand. Do you solemnly swear that the testimony that you give before this committee in the matter now under discussion to be the truth, the whole truth, and nothing but the truth, so help you God?

Mr. BENHAM. I do.

The CHAIRMAN. Have a seat, Professor Benham, and will you give us your background, sir?

STATEMENT OF THOMAS A. BENHAM, PROFESSOR OF PHYSICS, HAVERFORD COLLEGE, HAVERFORD, PA.

Mr. BENHAM. Well, my name is Thomas A. Benham. I am professor of physics at Haverford College in Haverford, Pa., and I have been interested in satellite tracking ever since Sputnik No. 1 was put into orbit on October 4, 1957.

I began at that time eagerly working on tracking by Doppler shift methods, the progress of the various satellites, the Sputniks 1 and 2, to begin with, and then the American satellites which came long afterward.

I have fairly good equipment at home—good receiving equipment, good recording equipment, good antenna equipment, and I have been following the progress not only from the point of view of tracking but from the theoretical point of view.

I have visited the Blossom Point Tracking Station, the Naval Research Computing Center, the Aberdeen Proving Ground Ballistic Research Laboratory Tracking Station. I have finally become officially part of the NASA through Dr. Hagen's efforts and am now on the Vanguard monitoring system.

I think, briefly, that gives a summary of what my interests are in the matter.

The CHAIRMAN. Thank you.

Now, what type of equipment do you have? You say you have very good equipment?

Mr. BENHAM. Well, very good equipment despite having gotten it together by hook or crook, to put the matter bluntly, going around to industry and asking them for assistance and getting a little money here and there where I could to buy things.

I have a good communications receiver which is used as a basic unit—that is the one on which I receive the Russian satellites directly with no additional equipment except an antenna. But for the International Geophysical Year program, which was transmitting on 108 megacycles instead of 20 megacycles, I have a special converter, an antenna system that is mounted at some distance from the receiving equipment where there is a fairly quiet location.

The antenna is a good one for a small installation. It has complete flexibility. It can be rotated any direction in the horizontal or vertical plane.

I have tape-recording equipment and standard laboratory equipment filters and audio oscillators, generators, and the like for comparing signals and analyzing signals, drawing graphs of Doppler curves, and things of that sort.

The CHAIRMAN. All right, sir.

Will you proceed with your statement?

Mr. BENHAM. Well, I am not quite sure how I am to proceed because I am not quite sure what statement I am to make.

Mr. ANFUSO. Mr. Chairman.

The CHAIRMAN. Well, we wanted to get what information you have in reference to lunik.

Mr. BENHAM. All right.

Now, if you have a recorder here which I believe you were going to have——

The CHAIRMAN. We have a recorder. Is it operating?

Mr. BENHAM. Can the recorder be brought over here so that I can run it?

The CHAIRMAN. Certainly.

Mr. BENHAM. Now, I think I am ready.

The CHAIRMAN. All right, proceed.

Mr. BENHAM. On January 2, 1959, the Russians announced at 6 o'clock eastern standard time, on Moscow Radio, that the lunik had been launched, that it had achieved cosmic speed, and that it had gone into orbit on its way to the moon. I think was the way they worded it.

They announced that in 1 hour's time it would be over the island of Madagascar, so I looked up on the map to see where that was and that was at 45° longitude.

I calculated that at approximately 3 or 4 o'clock in the morning, our time, on January 3, the rocket would be over this longitude, 75° W.

It would take approximately 8 hours for the earth to turn enough to get the rocket over this part of the earth.

So, I got up at that time and listened and I heard these signals [playing recording of signals].

Now those signals, as it turned out, are not the correct ones, but they matched fairly closely some of the releases that were made in the news shortly afterward, so I thought I had something. But I looked again the next morning—or rather, a week later I looked—after the announcement in the news that they were signals like that—I looked again a week later and they were still there.

So, I came to the conclusion that they were not correct.

Although they did match some of the announced signals, I felt they were not correct.

Then, at 4:30 that morning, I picked up Sputnik No. 3 as it was going by and thought I would put that into this recording that I am presenting now to you so that you can see that we feel as though we know what we are doing in that we were able to pick up the Sputnik No. 3 when we knew it was there and on the right frequency. [The recording is played.]

Now, that is the signal Sputnik 3 has been sending ever since it was launched on May 14, 1958.

At that time, that is, at 4:30 on January 3, a.m., I did not hear anything on 20 megacycles.

By the way, I did not say that the Russians announced in their 6 o'clock broadcast, our time, that the frequencies that were being used were 19.993, 19.995, and 19.997 megacycles and 183.6 megacycles.

Now, at that time I did not have any equipment satisfactory for looking on 183.6 megacycles, so I have nothing to report on that frequency; but on 19.995 megacycles that morning I heard one very brief signal that was in the right place; but it was very brief and unconvincing.

[The recording was played at this point.]

Mr. BENHAM. I am not used to this machine so I cannot hit things exactly right.

I will get used to this in a minute and I will be all right.

Now there; that signal is all there is.

I will do it again [playing recording]. That is all I heard in the hour that I listened.

Now, that is not a very convincing signal for a lunik to be sending back at 19.995 megacycles, especially since this was the first time that it was above our horizon.

As far as I can figure out, the lunik was launched at 11:41 eastern standard time—that is, Friday morning, just before noon on January 2—from somewhere in Russia and fired eastward the same as ours were and are.

It went over the horizon from their point of view and it stayed over our horizon until 3 o'clock or so the next morning, so that I do not believe that rocket was very far away yet—maybe 100,000 miles or so—and it should have been sending back a very strong signal still in terms of what was broadcast, as I will show you in a moment.

Now, the next thing I have is put on here for comparison purposes because, in Mr. Mallan's article, he remarked that the Russian satellite signals were weak and tenuous and undecisive, whereas the American signals were strong and clear.

I would like to present the following set of signals to listen to [playing recording]. Now, this is the signal repeated that I got that morning, January 3.

Now, this is the first day, May 14, 1958. [Playing recording.]

This was last night. [Playing recording.]

Now, there is nothing very faint or vague about those signals.

Now, the next thing that I have on the tape is a transcription of radio Moscow 7 p.m., our time, broadcast—that is, 7 p.m. our time.

I thought maybe you would like to hear that. It was 3 o'clock in the morning their time.

[Recording:]

This is Radio Moscow and here is the news. The U.S.S.R. today will launch a rocket toward the moon. It has entered its trajectory according to program and first reports indicate that the last stage has reached the required second cosmic speed. About 7 a.m. Moscow time, on January 4, the rocket will reach the vicinity of the moon. The last step of the cosmic ship weighs about 3,239 pounds without the fuel. It contains instruments for scientific research to gage the magnetic field of the moon, the intensity and variations of intensity in cosmic rays outside the earth's magnetic field, protons in cosmic radiation, radioactivity of the moon, the distribution of various nuclei in cosmic rays, the component of gas in cosmic matter, the radiation of the sun and meteor particles.

The communications with earth consist of a radiotransmitter operating on 19.997 and 19.995 megacycles sending reports of a duration of eight-tenths to one and six-tenths seconds.

A radio report operating on 19.993 cycles sending reports every half to nine-tenths second. A radio transmitter operated on 183.6 megacycles to measure the parameters of the rocket movements and telemetering scientific information to earth.

The rocket also has special equipment for producing a sodium cloud to serve as an artificial comet.

It will be possible to observe and photograph the artificial comet through a filter separating the sodium sector.

The manmade moon rocket will be photographable at 8:50 a.m., Moscow time, just short of 1 hour from now. Aboard is a pennant bearing the insignia of the Soviet Union and inscribed "Union of Soviet Socialist Republics, 1959." The total weight of the container is about 795 pounds.

Stations in different parts of the U.S.S.R. will track this interplanetary flight which will demonstrate the high level of Soviet rocketry and the advanced achievement of Soviet science.

The organizations which participated in building and launching the moon rocket have dedicated their achievement to the 21st Congress of the Soviet Union which Congress——

Now you can hear the signal of the rocket on its interplanetary flight through space toward the moon.

[End of recording.]

Mr. BENHAM. Now, those signals that were broadcast consist of two tones, a low tone and a high tone, about 2,000 cycles apart. The cloud, by the way, the sodium cloud by the way, that they were to have released in 57 minutes from the time of that broadcast was not photographed as far as I know.

The head of the visual tracking system for Russian satellites and space probes, Dr. Masevich, was at Haverford College on January 7, quite a propitious moment for her to arrive. At that time she gave a talk on the tracking situation in Russia, how they track their satellites by optical methods, and during that talk she said that the cloud, the sodium cloud put up by that rocket, was not visible from Russia because of a heavy overcast and the only photograph that she knew of at the time was taken by someone in Scotland.

I believe the photograph that a Scottish astronomer took has really been doubted as to whether or not it was a picture of the sodium cloud.

On February 9, Mr. John Puzzini of the National Security Agency called on me to see what I had in the way of signals from the lunik. At that time he presented me with a tape recording that was made by one of our military installations on Hawaii at 12:01 noon, Eastern standard time, 20 minutes after launching time.

So I will play an excerpt of that tape that he gave me. There is some confusion in it because there is what he told me was a Japanese timing signal background similar to the timing signal that we use on WWV. That is right out here—here in Beltsville, Md. But the Russian satellite signal is very clear, even though it is confused with this other signal and you can tell that the pattern is the same as the pattern that was broadcast in the broadcast we just heard.

The CHAIRMAN. Is that the lunik signal?

Mr. BENHAM. That is what he told me. Yes, sir.

Now, you see there are two different frequencies about 2,000 cycles apart again.

[Recording.]

Mr. BENHAM. They sound very similar.

Mr. ANFUSO. Mr. Chairman.

The CHAIRMAN. Is that supposed to be the lunik signal that you referred to?

Mr. BENHAM. Yes, according to Mr. Puzzini that was recorded in Hawaii 20 minutes after launching time.

The CHAIRMAN. And you transferred it to your tape?

Mr. BENHAM. Yes.

Mr. MILLER. Which is the lunik—the high-pitched one, or the low-pitched one?

Mr. BENHAM. Both of them.

It shifts back and forth.

The CHAIRMAN. Lunik was sending out two signals at that time?

Mr. BENHAM. Yes, first one and then the other.

Mr. ANFUSO. Mr. Chairman.

The CHAIRMAN. Mr. Anfuso.

Mr. ANFUSO. Do you disagree with the finding of Dr. Puzzini?

Mr. BENHAM. Heavens, no. I am only presenting what I have.

I have no opinion as to whether the Russian lunik was a success or not. All I am saying is I have no signals that would indicate it was there. But I do not know whether it was a success or not, because I do not have the best equipment in the world, and I did not have anything on 183.6.

I do know that I heard nothing on 20 megacycles, let us call it 20 megacycles instead of 19.997 or something.

I heard nothing on 20 megacycles that would indicate that it was there, but that does not mean that it was not a success.

The CHAIRMAN. Just proceed, sir.

Mr. BENHAM. Now, after that, a Mr. Barry Miller, who is on the staff of Fairchild Publications that puts out Electronic News, contacted me on about January 15 and asked me what I thought of it. I told him essentially what I just said, that I did not hear anything, but that did not prove that it did not exist.

I think it is fairly clear that there was no transmission on 20 megacycles that first day, that is 3 a.m. our time, January 3, because I feel, if there had been a signal there then on 20 megacycles, I would have heard it. But I heard nothing resembling the Russian broadcast or the signals that Mr. Puzzini gave me. Mr. Miller of Electronic News got in touch with me and asked me would I analyze tapes sent in by others, because he had word from a Mr. Filer, of the Sanitary Corp. in Chicago, and a Mr. Jones, of the Standard Oil of Ohio in Cleveland, that they had recorded signals on 20 megacycles. They would be interested in having me compare them with what I had and lending my experience in listening to signals to see whether I thought there was anything there.

Now this next recording is fairly long—maybe 5 minutes—and, therefore, I will tell you what is going on as we proceed.

The CHAIRMAN. All right, if you will.

Mr. BENHAM. Now this was sent by Mr. Filer. [Recording:]

Signals from Richard Filer of Chicago, Ill.; moon shot from Russia, 8:30 to 9:30, January 3, 1959.

Now there was a long dash there—you might not have heard it. There is another one, it is weak. There is another one. Now there is some more. There it is—a little short dash. There it is better. Now there is static in there. [Continuing with recording.]

There was a funny one. He was tuning the receiver when it did that.

There is another one. That was a long one. [Recording:]

This long dash was at 9:17 a.m., January 3.

That is Mr. Filer. I left it on there. [Continuing to play recording.] That is a short one. [Recording:]

That was 7:12:30 to 7:14 a.m

Now I will stop it a second because this next thing is what makes it all seem doubtful to me, that they were actually signals from lunik.

The next signals that we are going to hear—there is no doubt in the mind of anyone who is experienced in listening to such signals—that these are signals from a teletype station sending a teletype message from somewhere to somewhere else.

[Recording:]

Mr. MILLER. Could they have been sent from an oblique lunik?

Mr. BENHAM. Not in my opinion, because anything that would send any signals as complicated as that would require some very, very involved electronics, plus the fact that there had been nothing but short dashes and then all of a sudden this complicated signal starts to come in.

Now, in a minute he tunes the receiver while he has it on and when he does that, there is no doubt of it in my mind, that it is a teletype.

Mr. MILLER. You speak of 20 megacycles and then you gave us 19.993, 19.995, and 19.997. If you were tuned in on 20 megacycles, would you pick up those others?

Mr. BENHAM. Yes. The receiver that Mr. Filer was using has a very sharp filter on it, as they call it, so it is possible to separate frequencies that are very close together.

So does my receiver, so I can do it the same as I will show you in a minute, but it is possible to receive signals that are very close together on one of these good receivers.

Now here is where he tunes it [recording]. There it is. I think there is one more. I think I say something about it. [Recording:]

The following was recorded at about 9:50 a.m., Sunday morning, January 4. Note the W.W.V. signals in the background.

That was a comment I put on there. This is still Mr. Filer's material. [Continues to play recording.]

There was a dash. There is another one. He tuned the receiver there. That is what made the pitch change.

But the important thing about this particular recording is that you can hear W.W.V. second ticks in the background. It is not very clear but it indicates he was on the right frequency.

So they were the signals sent by Mr. Filer.

Now, Mr. Jones sent me the following signals. As far as I am concerned, he recorded them at a time that they could not exist.

He told me that he recorded these between 8:30 and 9 p.m. on January 3 and at that time the lunik would be over the horizon, if we did hear it, it would be because it had come through the ionosphere reflected off the ground, reflected from the ionosphere generally and come back to us that way by a multiple reflection.

Now, for a signal that was as far away as lunik would be by that time, maybe 150,000 miles or something, for it to be that far away and for it to have bounced around that many times makes it highly doubtful that these signals came from it.

[Recording:]

The following signals were recorded from a tape sent by Mr. Jones of Standard Oil of Ohio, Cleveland, Ohio, recorded between 4 and 5 a.m.

Mr. BENHAM. The time is wrong on the tape; it was between 8:30 and 9 p.m. on January 3.

(Playing recording.)

Mr. BENHAM. Now, that is a fairly strong signal.

Mr. MILLER. Who are these people that have given you these, Mr. Jones and these others?

Are they officials or amateurs?

Mr. BENHAM. They are private. Standard Oil owns the one Mr. Jones is using and the other company owns the one Mr. Filer is using. I have nothing official from any of the Government tracking stations on lunik.

I have corresponded or written to Dr. Pickering and Dr. Dryden, and to NASA headquarters and other places to try to get what might be termed an official signal but I have not been successful.

Mr. MILLER. So then, all of these signals which were played for us now were signals picked up from nonofficial sources, most of whom are, some of them at least, so-called hams in this business?

Mr. BENHAM. Yes; except the fact that Mr. Filer and Mr. Jones were using equipment that was under the research departments of Standard Oil of Ohio and the Chicago firm, which makes it a little bit more than just a ham.

Mr. MILLER. I appreciate that. The point I wanted to bring out was that that is the source.

Mr. BENHAM. That is right; it is not official.

Mr. MILLER. It is unofficial sources.

Mr. BENHAM. I find it a bit frustrating that I cannot get signals from an official source.

The CHAIRMAN. Are you satisfied that the signals you are getting there, that the ones you indicated, were coming from lunik?

Mr. BENHAM. I am satisfied that the Russian broadcast was a transmission from lunik.

Where Lunik was when they recorded it, I would not know. I am satisfied that Mr. Puzzini gave me an honest recording of signals recorded in Hawaii.

The CHAIRMAN. He is a pretty good man.

Mr. BENHAM. I do not know him other than the time he came to see me, so I do not know anything about this authenticity.

The CHAIRMAN. Well, your position is you do not doubt the existence of Lunik but you are not prepared to say whether it was successful or not.

Mr. BENHAM. That is right, I do not doubt that they launched something but what happened to it, I do not know.

Mr. MILLER. We have launched a lot of rockets that we have not heard from.

Mr. BENHAM. Indeed we have and I have recordings of many of them that went up and came down again.

Mr. MILLER. Have you any thoughts on the Russian equipment? Is it poor, as poor as Mr. Mallan makes it out to be?

Mr. BENHAM. I have opinions. My opinion is that they must have something fairly decent because the satellites that have been put up by the Russians are genuine satellites.

He spoke in his article, quoting Dr. Leavit, of the Franklin Institute, that the orbit was unsteady or would be high or something of that nature, indicating they just threw it up there like you might throw a stone and it happened to stick. I do not think it is quite as haphazard as that.

The first sputnik stayed up 3 months. Our Discoverer stayed up a couple of weeks. Their Sputnik No. 2 stayed up 4 or 5 months. Our Discoverer 2 stayed up 1 week.

Our Atlas stayed up only 2 or 3 weeks. So, I do not think that the speed with which they come down has anything to do with how clever they were in getting it up there.

On the other hand, the Sputnik 3 which weighed 3,000 pounds roughly, was put up on May 14, 1958, and it is still there.

Its perigee is down around 135 miles now. It has dropped from about 190, I believe it was—something of that nature—which indicates that it is coming down slowly and it will come down eventually, maybe within another year.

I am not sure, I have made some calculations that would indicate that it might stay up another year, but I do not believe that the speeds with which the satellites come down has anything to do with the cleverness of the device which puts them up there, because some of ours have come down in a hurry.

I do not believe that the Air Force intended the Discoverer satellites to come down as soon as they did, because they were after information about the north-south orbit that they went to great pains to put it into.

So, I do not believe that that is a criterion. Now there is one piece——

Mr. SISK. Mr. Chairman, could I ask the doctor one question, since he is discussing those Russian satellites?

Now, some mention was made a little while ago with reference to the weight of the present Russian satellite of some 3,000 pounds. Now, how authentic do you believe that approximate weight to be?

Mr. BENHAM. I do not know about this one but I know about the other two.

Sputnik No. 1 was alleged to weigh 184 pounds.

When it finally broke up in January 1958, it broke up into some small pieces and Dr. Krause who is an authority on radiation and antenna theory—I believe he is associated with the Ohio State University, if I remember correctly—had a special program in which he was tracking the satellite after its radio had gone out of existence by what is known as passive tracking. You take a signal from some other station, he was using WWV—on 20 megacycles, and pick up the reflection of that signal off the pieces or off the satellites or off the trail that it leaves as it goes through the upper atmosphere.

He tracked that completely until it was broken up into, I think it was, five pieces; it might have been three. I do not want to be quoted on the number of pieces, but it was several pieces, and from the way in which those pieces behaved, calculations were made backwards to find out how much they estimated it weighed and they came out with something close to 184 pounds.

They did the same thing with Sputnik II, the "dog" satellite. When it came down they calculated backwards from what happened to it and came up with the idea that it weighed somewhere in the neighborhood of 1,100 pounds and the Russians quoted 1,168, I believe.

So, I believe that those two have been qualified as being close to the right weight. Now, of course, Sputnik III has not come down yet, and it isn't until it comes down and breaks up that they can tell as far as I know.

Mr. SISK. Thank you.

Mr. MILLER. Mr. Chairman——

Doctor, of course the thing that concerns us all, is the statement made here today which would indicate that we have sort of lost our heads, that we have been hoodwinked by Russia. Do you think that we are ahead in this race, or the Russians are ahead, or is there very much difference between our efforts and their efforts?

Mr. BENHAM. I think we are ahead but in a different way. I think that the Russians do what they do by a more or less "brute force" method. We have visitors pass through Haverford, physicists from other institutions, and I have spoken to some who have been to Russia. They would corroborate Mr. Mallan's statements that the bulk of the equipment is enormous compared with what we send up, but I do not believe that that has anything to do with whether they send it up or not.

It may be that they use large vacuum tubes and heavy batteries and whatnot, but I do not believe that this has anything to do with whether they have successfully launched something or not.

If they can do it with heavy equipment the way they claim they have done and the way the American scientists have corroborated they have done, more power to them; as a matter of fact, we have not been able to send up anything quite so elaborate in the way of weight. Now, the Atlas satellite belies that to a certain extent. I think maybe on that one we are about even because that weighed about 8,700 pounds whereas the Sputnik III plus its carrier weighed something like 10,000, so that may be close, but I do not believe whether it is miniaturized or whether it is not has anything to do with whether they are successful or not.

One of these physicists I was thinking of today did say they did not show any miniaturization at all, in their displays of Sputnik III and Sputnik I and Sputnik II but I say, "so what."

Mr. MILLER. They got them up there and "so what" is right.

The CHAIRMAN. Do you have more signals now to show us?

Mr. BENHAM. Just one on this tape.

The CHAIRMAN. All right. We had better finish with them because we are going to have to adjourn in a few minutes.

Mr. BENHAM. There is one more comment on that while you are on the subject of miniaturization.

I asked Dr. Masevich when she was at Haverford why they did not miniaturize the equipment in the satellites.

She said why experiment with two things at once, miniaturization and rockets?

That was her answer.

Now this last signal which I have on here, on February 11, between 4:45 and 5:05 eastern standard time, that is p.m., I thought it would be fun to look on 20 megacycles as carefully as I would have looked if I had thought Lunik was there.

Now, of course, on February 11, we all know that Lunik was not there then, so I looked as carefully as though it were there, and I picked up all kinds of strange things that could be interpreted as Lunik signals.

[Recording:]

Recording of the signals recorded by T. A. Benham at Haverford College on 19.999 megacycles February 11, 1959, between 4:45 and 5:05 p.m. eastern standard time.

Mr. BENHAM. Do you hear that?

That sounds very similar to Mr. Filer.

There is another one.

Now, the only point in doing that is to show even as of February 11 there were signals there that sounded familiar, judging from Mr. Filer's recordings.

Now, this is not to say that Mr. Filer did not record the Lunik. This is only to say that I doubt it because the same kind of signals were there on other occasions.

Well, that is enough of that.

The CHAIRMAN. Professor, the bell has rung and we have to vote over on the floor of the House.

Summarizing what you have to say, though, you do not have any doubt about the existence of Lunik in really being fired aloft and its going into space.

The thing in your mind that is doubtful is whether it was successful in bypassing the moon and going into orbit around the sun?

Mr. BENHAM. I do not even have any questions about it. All I say is I have nothing in my files to prove that it was a success, that it reached the moon. I certainly wish that someone would make it possible for me to have a recording. If the authorities say that it was a success, I wish they would present me with some recorded evidence of it to go along with what we have.

The CHAIRMAN. You do not have any doubt about the success of Sputnik II and III, do you?

Mr. BENHAM. No; that was there and very definite.

The CHAIRMAN. If they were successful with Lunik, it would be an extension of I and II?

Mr. BENHAM. Yes, and III.

The CHAIRMAN. It would just be further progress along the same line or following the same path they had already started on?

Mr. BENHAM. That is right. Now, of course, what they may have done was to shoot the rocket up there and decide what it was going to do after it went, and this corroborates a little bit what Mr. Mallan stated, except that the Army did the same thing.

They aimed a rocket in the vicinity of the moon, but they left a loophole, because they said it might go into orbit around the sun, which is what it did. They did not hit the moon with it.

The CHAIRMAN. Well, we certainly thank you, Professor Benham, for coming here and helping us. That is the first time I have heard signals of that character.

Mr. HALL. Mr. Chairman.

The CHAIRMAN. Mr. Hall.

Mr. HALL. Doctor, you started to make a statement a moment ago—"I wish that somebody in this country would"—and you did not complete it.

Mr. BENHAM. I guess I was reiterating my wish that someone would give me official signals proving it was a success, if it was, because I would like to have them.

Do you want this tape for the file?

The CHAIRMAN. Yes; we will put that in the file, if there is no objection.

Now, Professor, another statement. Could you supply for the record exactly what type of equipment you are using there? You just put that in the record, if you will.

Mr. BENHAM. For the sputnik signals?

The CHAIRMAN. Yes; you referred to it in a general way; but if you could use the more technical nomenclature, we would appreciate it.

Mr. BENHAM. I have been using a Hammerlind H.Q. 120 Q. receiver for the sputnik signals with a half-wave dipole antenna oriented north-south and a Revere tape recorder, model T. 700.

The CHAIRMAN. Thank you very much.

Now we will conclude this portion of the hearing, and, as I announced earlier, I am going to refer this particular inquiry regarding the article in True magazine entitled "The Big Red Lie," by Lloyd Mallan, to a subcommittee to be headed by Mr. Anfuso, of New York, with Mr. Wolf, Mr. Karth, Mr. King, Mr. Osmers, and Mr. Van Pelt as members of it.

The inquiry regarding Russian science progress will continue tomorrow morning on the general program, whereas this particular inquiry will be turned over to the subcommittee.

The committee will adjourn until tomorrow morning at 10 o'clock.

(Whereupon, at 12:30 p.m., the committee adjourned until 10 a.m. Tuesday, May 12, 1959.)

SOVIET SPACE TECHNOLOGY

TUESDAY, MAY 12, 1959

HOUSE OF REPRESENTATIVES,
COMMITTEE ON SCIENCE AND ASTRONAUTICS,
Washington, D.C.

The committee met at 10 a.m. in room 214-B, New House Office Building, Washington, D.C., Hon. Overton Brooks (chairman) presiding.

The CHAIRMAN. The committee will please come to order.

We are very glad to have you, Dr. Dryden. Some of this has to be in executive session. Dr. Dryden may have to refuse to answer some of the questions in open session. I know the press will understand it. Doctor, if you will proceed.

STATEMENTS OF DR. HUGH DRYDEN, DEPUTY ADMINISTRATOR, NASA; DR. HOMER E. NEWELL, JR., ASSISTANT DIRECTOR, SPACE SCIENCES, OFFICE OF SPACE FLIGHT DEVELOPMENT; AND DR. WILLIAM H. PICKERING, JET PROPULSION LABORATORY, CALIFORNIA INSTITUTE OF TECHNOLOGY

Dr. DRYDEN. I have no general opening statement. We are here to answer your questions. I think we might begin by having Dr. Pickering describe the tracking operations on lunik, or mechta, as the Russians call it. Then we might proceed from there with questions.

The CHAIRMAN. Lunik is the U.S. name, and mechta is the Russian name.

Dr. Pickering, you can proceed with either one.

Dr. PICKERING. I will describe, then, the exercise, if you like, which we went through on 2 nights, the Friday and Saturday nights after the launching of lunik.

We received first word in the afternoon of Friday in Pasadena that lunik had been launched. We decided we would attempt to track the vehicle from our Goldstone tracking station. It is a site on the desert approximately 150 miles from Pasadena, Calif.

We received some information on the frequencies which the Russians had reported were being used. We found that we had no equipment actually installed which would operate on those frequencies. So we checked equipment from the laboratory and collected receivers and picked up some antennas which could be used at the focus of the "Big Dish."

Then we went out to Goldstone. Most of the night was occupied in installing these antennas and this receiving equipment on the dish itself.

The receivers are mounted physically on this structure. This is a large structural steel-tracking antenna, 85 feet in diameter. The receivers have to be mounted upon the dish itself. The antenna is put at the focus of this dish.

Mr. MITCHELL. Did you have any advance notice that lunik might be launched?

Dr. PICKERING. No advance notice of the launching or frequency.

Mr. MITCHELL. No notice until after the launching?

Dr. PICKERING. The first announcement was over the radio. By early Saturday morning we had the equipment working after a fashion. So we proceeded to try to find the object with the antenna looking in the general vicinity of the moon.

We tried for several hours without success. During this time we were receiving reports of various kinds from people who perhaps thought they had heard the signals. We investigated many of these reports. Many of them turned out to be false.

On Saturday morning, then, was the first chance that we had to really see how well we had installed the equipment. We did some calibration on Saturday morning. We discovered that the equipment was not working very well, and we proceeded to improve the equipment during the afternoon of Saturday. Then on Saturday night we were ready again at moonrise, which was at about 2:30 a.m. Sunday morning. We decided this time to concentrate on the one frequency of 183.6 megacycles which the Russians had announced was the frequency being used.

At about 3 a.m. we first received signals which we believe came from lunik. Let me describe the procedure, please.

We had the antenna point slightly to the west of the moon. According to the Russian report, lunik should have passed the moon before we were tracking early Sunday morning.

In that event, as seen from Pasadena, the object should have been slightly to the west of the moon. We started looking to the west of the moon. The tracking antenna was placed in a search mode, which means it was scanning around in a spiral.

It would work out to perhaps a 10° spiral and then start over again working out, and centering this entirely somewhat to the west of the moon. When the people on the dish—let us see—the receiver is physically up on this steel structure, and the operators are up on the structure likewise. The controls for the dish which are putting it in this tracking mode are down at a control house on the ground. When the men up on the antenna report receiving a signal, the dish would be stopped searching and perhaps moved a little bit to see if the signal could be picked up a little more strongly. And then when it was reported that they were receiving successfully, on the ground we would push a button which would record the direction in which the dish was pointing at that instant of time. This exercise was continued for several hours, and the results I think are summarized in this chart.

The results can be plotted in a number of ways. What we have done here is plot Greenwich mean time horizontally. In the other direction we have plotted the angular position, the so-called hour angle using the astronomer's terminology here. This just gives us the angle, if you like, from the meridian plane or from the horizon the angle that the dish was pointing at. As time goes on, the angle is increasing from about 300° to about 350°.

This means that the dish is moving to the west. The green line shows the path of the moon during this period. The red line with the dots on it, the dots indicate the directions in which the dish was pointing at the time that the observers reported receiving signals.

You notice there is one time at about 11:30 Greenwich time, and then there is a group of signals around 1300 Greenwich time for about a half hour there, which were being received quite well. Then again in the early morning in Pasadena at about 1500 Greenwich time, which is about 7 a.m. in Pasadena, signals were received. There were three periods when signals were received satisfactorily.

I point out again the plot here, giving the direction in which the dish was pointing was attained by the man on the receiver saying "I am receiving a good signal," the man in the control house pushing a button and pointing the dish. There is no correlation, as it were, between the moon's direction—in other words, the man on the dish does not know where the dish happens to be pointing when he says "signal." It is not a subjective relationship here is what I want to point out.

There are two different observers involved, and the data obtained in this fashion, as I say.

This shows that data then which first of all, says we were apparently tracking an object which was moving away from the moon to the west, because it is moving a little further west, a little more rapidly than the moon. This is what would be expected of an object which had gone beyond the moon and was going on out into space. The moon is traveling around the earth, so the moon is moving toward the east, the object going out into space is going more or less straight out.

One observes then two things, first of all, that the data, although it exists at only three times, the data is consistent. Secondly, that it is consistent with an object which is past the moon.

I have one other plot of this data which I believe——

The CHAIRMAN. How far out now is that object?

Dr. PICKERING. About a quarter of a million miles.

The CHAIRMAN. Is it unusual that your instruments would be able to track something at that distance?

Dr. PICKERING. No; on Pioneer 4 we tracked out to considerably greater distance.

The CHAIRMAN. That is not strange at all, the ability of your instruments to track an object moving at that distance?

Dr. PICKERING. No.

The CHAIRMAN. Is there any doubt in your mind that the object you were tracking was this lunik or Mechta as you call it?

Dr. PICKERING. There is no doubt in my mind.

The CHAIRMAN. You feel like the information obtained over the radio from the Russians and all that was correlated corresponded exactly with the chart that you now have before the committee?

Dr. PICKERING. Yes, sir. I would like to present this other chart as additional evidence.

Mr. KING. You say that the results that you have on the chart are consistent with lunik's being where the Russians claimed it was. Are those facts consistent with any other theory?

Dr. PICKERING. I do not know what it would be.

48488—59——5

Mr. KING. I am asking you: Is it conceivable, in other words, that lunik could be placed someplace else and you still reach that conclusion?

Dr. PICKERING. No; if it was a satellite, it would be moving much more rapidly across the sky. There is one other possibility that perhaps one thinks of. If you want to explore possibilities here, that would be the case of reflections from the moon. It is well known if you take a sensitive radio receiver and point it at the moon you can hear signals which are reflected—signals which have originated on the earth and bounce off the moon and come back again.

I think here we are showing an object moving away from the moon which is significant. It is not just pointed at the moon.

Mr. KING. This would be inconsistent with reflections from the moon? They would not give the same results you have there.

Dr. PICKERING. That is right.

Mr. BASS. You did not pick up the signals until after lunik passed the moon.

Dr. PICKERING. On Friday night lunik was between us and the moon. But we discovered on Saturday morning that our equipment was not working very well. We had no success Friday night.

On Saturday night we received these signals starting at 11:30, which is 3:30 a.m. Pacific time, and moonrise was approximately at 2:30 a.m. This was almost as early as we had an opportunity to track.

Mr. BASS. Why was there an interruption of signals received? In other words, after you had first picked up these signals, why didn't you receive them continuously?

Dr. PICKERING. My only answer is that the signals were very weak. We were sort of working on the limit of the sensitivity of the equipment. And little changes could move the signals in or out. The time around 13 o'clock was when the signals were strongest; that was the time we received them for quite a long period of time.

At the other times they were weaker——

The CHAIRMAN. How far out did you trace that object?

Dr. PICKERING. This was all, sir.

In other words, we did not try again the following night.

The CHAIRMAN. 250,000 miles is as far as you traced it?

Dr. PICKERING. Approximately.

The CHAIRMAN. You do not know whether it went into the sun orbit or stayed in the moon orbit or what happened?

Dr. PICKERING. If it has passed the moon, as the Russians reported and as we believe this data shows, then it must go into an orbit around the sun. There is nothing else for it to do.

Mr. KARTH. Yesterday it was suggested that for one reason or another you may have had your signals crossed and you were getting signals from Jupiter rather than from lunik. Is there any possibility that you had erred in whom you were getting signals from?

Dr. PICKERING. I would like to draw your attention to this other curve. Could we pass these out?

Mr. KARTH. Could you give me my answer?

Dr. PICKERING. It is on here. I will have to explain this. It is a little bit more complex but let me try to explain it. I have plotted the two lines—declination and right ascension on the position of the lunik at various times on the 3d and 4th of January.

You notice there is a jump in the data on right ascension. This is just taking the data which was received from the Russians via the Smithsonian Institution.

The reason for that jump, I assume, is that they corrected their data as they were analyzing it.

On the righthand side I have a series of dots on both the declination chart and two dots on the right ascension chart.

Those dots represent the data that we received at Goldstone. I submit that what we have here, then, is evidence that the declination readings, the declinations according roughly to latitude of the object, that these readings which we received are all within about half a degree or a degree of the value as given to us by the Russians.

The right ascension values spread over about a total of 4°, plus or minus 2°, centered on the value given to us by the Russians. Therefore, the experimental data received at Goldstone agrees very well with this data which we received sometime later from the Russians on the position of the object. The Russians' data, I will point out, was received some days after the Goldstone data had been received.

Mr. KARTH. There is no possibility that Jupiter could have given that?

Dr. PICKERING. No; Jupiter is indicated on the chart as 230° which is way up here. The points here show no relation to Jupiter, but do show a very nice agreement with the data which the Russians published on the position of their lunik.

Mr. KARTH. Anyone understanding these fundamentals would not make the mistake of suggesting that these signals might have come from Jupiter; is that correct?

Dr. PICKERING. That is correct.

Also, there is the question of the nature of the signals. The signals from Jupiter, I understand, are signals similar to static from thunderstorms. The signals as reported by the men on the receivers had a characteristic which was somewhat like the sputnik signal. That is, it was a signal which was fluctuating periodically.

Mr. KARTH. He further suggested that these signals were very weak and indecisive and, as a result of that a mistake could have been made.

Dr. PICKERING. The signals were very weak. We were not able, as the chart shows, to maintain the signals long enough to get continuous tracking or to get very precise tracking.

Mr. KARTH. Despite their weakness, there was not the possibility of such an error as he suggested?

Dr. PICKERING. I believe not. For the two reasons that we were pointed about 15° away from Jupiter and that the signals from Jupiter would have been of a staticlike nature where these were signals which were manmade signals.

Mr. KARTH. Is there anyone who is tracking Jupiter, anyone else tracking Jupiter at precisely the same time and therefore had its whereabouts charted?

Dr. PICKERING. The whereabouts we know from astronomy. Whether there were any signals from Jupiter at that time, I do not know.

The CHAIRMAN. Do I understand you to say that Jupiter puts out signals?

Dr. PICKERING. The radio astronomers say they get signals from Jupiter which have the characteristic of thunderstorms.

The CHAIRMAN. Do all planets put out signals?

Dr. PICKERING. Venus and Mars also put out signals.

The CHAIRMAN. Are their signals different from the——

Dr. PICKERING. They are noiselike signals. Perhaps Dr. Newell would like to comment on it.

The CHAIRMAN. If you received a signal, could you say whether it was from Jupiter, Mars, or Venus?

Dr. NEWELL. You would know from which planet it came by the direction in which your antenna was pointing.

The character of the signal is of two types, one what we call thermal noise which is an emission, a radio emission from hot gas. The other is the lightninglike or staticlike noise that Dr. Pickering referred to on Jupiter. The signals that have been obtained from Venus and Mars have been of the thermal noise type and those from Jupiter——

The CHAIRMAN. You can identify the different planets. I think somebody called that the music of the spheres. It does not sound much like music from the way you describe it. You can identify the noises coming in by the location of your source—gas, is it—and also in some instances by the variation in noises.

Dr. NEWELL. Yes.

The CHAIRMAN. Is that noise different with manmade vehicles like lunik?

Is that to be expected?

Dr. NEWELL. It is quite different. It is a random affair and not uniformly modulated the way a manmade signal would be.

The CHAIRMAN. A manmade signal is more uniform, more modulated, and therefore you could not mistake it if you were ranging out in deep space? You could not mistake a manmade object out there by its signal from a sphere?

Dr. NEWELL. Normally not.

Mr. TEAGUE. Would you tell us generally what American scientists have been to Russia, what they were allowed to see——

Dr. PICKERING. I would like to refer that question to Dr. Newell, himself. I have not been there.

Dr. NEWELL. As to the first part of your question, what American scientists have been to Russia, in the last several years our scientists have been going to Russia to several conferences.

Last summer for example, there was a meeting of the International Committee for the International Geophysical Year, in which some scores of American scientists were present along with scientists from the rest of the world.

There was a meeting of the International Astronomical Union at which a large number of American scientists were present.

We got to visit—that is by "we" I mean the scientists, American scientists got to visit—a number of the astronomical and magnetic ionospheric observatories in the Soviet Union. We got to see the model of Sputnik III and Sputnik II which were on display in the fair, which is a continuing fair in Moscow.

We got to visit the adjuncts of the University of Moscow. However, we did have the impression that what we were shown was carefully controlled.

Mr. TEAGUE. What was your impression of the equipment that you saw? Is it outdated and old navy stuff, or is it equipment that you would like to have in your place?

Dr. NEWELL. It was good equipment. It was equipment of which no one had to be ashamed.

In fact, I have brought along a number of reprints of a paper which John W. Townsend and myself wrote up after this meeting. We would like to have you see these, and would like to read to you, if you wish, at least a few excerpts from the paper to illustrate what we saw.

Mr. TEAGUE. What does a scientist know when he looks at a model of sputnik? Can he look at a model and tell that it is something or is not something?

Dr. NEWELL. In the case of the Sputnik III we not only saw the model, but we saw much of the equipment laid out in open form on a bench nearby so we could look at the individual instruments.

One could tell from the way in which the wiring was done, the mechanical construction was worked out and so on, that it was very well done.

Mr. BASS. Dr. Pickering, I would like to know a little bit more about the nature of these signals that you received.

Were they a series of dots and dashes? Could you make anything out of them? To a layman's ear like myself, would they sound like a code?

Dr. PICKERING. The signals from sputnik are characteristically a beep-beep-beep, a series of dashes whose length changes and separation changes as the scientific measurements vary.

In the case of lunik, the signals instead of being turned on and off, instead of going "beep, beep, beep" were two different tones.

So, it was going be-ah-be-ah—that sort of thing. But it was still a case of signals which were fluctuating from one tone to another.

This was the description given to me by the men on the radio receiver at Goldstone when they came down off the receiver. They said it was similar to sputnik but differed because it was a two-tone modulation rather than an on-off modulation and the swift frequency was much more rapid than sputnik.

Sputnik goes about "beep-beep-beep-beep" and this was more like five times a second. That sort of thing; some days later we received a tape recording of a broadcast which was reportedly the signals from lunik.

This tape recording did show this two-tone modulation. The difference between the tape recording and the report from my men on the dish was only in the switching frequency.

The recording which Moscow radio broadcast was a switching frequency of perhaps once a second. And, as I say, the men on the dish reported it more like five times a second.

I would interpret this as meaning that the switching frequency was making a scientific measurement and the quantity had changed between the original Moscow broadcast and after it had passed the moon.

Mr. BASS. Were you able to glean any information from these signals or get any information from them?

Dr. PICKERING. No. The Russians have published a number of reports on this material, on the results, rather. The first publication was in Pravda of March 6, and we have just this morning received a translation of a publication in a scientific journal dated April 11. This paper in the scientific journal is more detailed than the paper in Pravda, although the same author or one of the authors is the same in both. Two of them are the same in both.

Both papers are concerned with the measurements of radiation on the lunik experiment. I might comment, incidentally, that in a general way the data presented here is quite similar to what we received on Pioneers III and IV.

Mr. MITCHELL. How do the signals you received at the same distance from Pioneer IV compare with these lunik signals?

Dr. PICKERING. The Pioneer IV signals were stronger. I think one must be pretty careful about making a comparison. The strength of the signals which you receive and useful signals will depend on the sensitivity of your receivers, on the power that is being transmitted, on the kind of transmitting antenna and on a number of factors.

So that, in other words, the fact that we received weak signals from lunik and relatively strong signals from pioneer does not necessarily mean that we had a stronger transmitter than they did. It could mean that we had a poorer receiver.

In fact, I know this is true because the receiver we used for the lunik experiment was something that was picked out of the laboratory and rushed out onto the desert and it was not a very good receiver.

Mr. MITCHELL. There has been some discussion of the other disturbances that you did confuse them as being signals from lunik, when actually they were other disturbances out in space giving a similar sound.

Dr. PICKERING. I presume this is referring again to signals from Jupiter or signals from radio stars of various sorts. Those signals are of a different type of noise altogether than what we received. In other words, there was no mistaking what we received. It had to be an artificial source. It was not a natural source. The natural sources have an entirely different character.

Mr. FULTON. There is no possibility that there might have been an echo from some other artificial source?

Dr. PICKERING. That is one thing that did concern me, that we were receiving echoes.

When we plotted data as shown on this chart, the data is definitely away from the moon and moving away from the moon. An echo, then, would not do this; it would be an echo from the moon. It is possible to receive echoes from the moon on almost any frequency, as a matter of fact.

Mr. FULTON. There is no chance that some other U.S. service might have been echoing some sort of a signal from the moon?

Dr. PICKERING. I believe not, sir, because of the fact that the signals definitely appeared to come from a point west of the moon and not from the moon itself. And the distance west was, I believe, larger than the sort of probable error in the measurement.

Mr. FULTON. Is there any correlation between the pattern of signals and any that the U.S. services might have used at any time?

Dr. PICKERING. I just do not know. I have not considered that. Let me say it is an unusual pattern—no; I will withdraw that. There are teletype signals that are not too different from this sort of thing. But teletype signals usually have an entirely different sound.

I do not believe this could have been a radio teletype.

Mr. FULTON. There is no possibility of a satellite being in orbit, and then on the other side of the world at some other point there would be an echo that would have caused this?

Dr. PICKERING. I do not believe so at this frequency. These round-the-world echoes must take place at a lower frequency, because you have to involve the ionosphere. At these megacycles the ionosphere has no effect on the signals.

Mr. FULTON. Wasn't there a case where they thought they had traced certain high altitude echoes and found that rather than another satellite or another space vehicle it was the echo from an already existing vehicle?

Dr. PICKERING. This was at 20 megacycles. But a very interesting effect, indeed, from the radio point of view, that diametrically opposite the satellite there was a sort of ghost satellite moving along.

Mr. FULTON. There is no possibility of that having been the case here?

Dr. PICKERING. No, because of the frequency. The radio frequency rules that out.

Mr. KING. I would like to ask two questions either of Dr. Pickering or Dr. Newell. I think I know the answer to the first question, but I will ask it, anyway.

Assuming you have the power or the thrust necessary to get a missile out of the gravitational pull of the earth and into orbit, does it still require skill to get it into orbit? Is your problem more than one of just raw strength and push? Is there also a great deal of skill and art in plotting it to get it into orbit?

Dr. PICKERING. Yes; the guidance problem is a difficult one. Of course, I will point out it depends on your objective.

If you just want to throw it away, the guidance problem is relatively simple.

If you want to throw it to the moon, the guidance problem is quite difficult.

Mr. KING. In this article "The Big Red Lie," the statement was made that the Russians admittedly had developed a lot of power and thrust and raw strength behind their missiles but that they had developed very little other than that. And that, in the field of guidance, and tracking and instrumentation and all these other subtleties, they were 10 or 15 years behind us.

My question is based upon what you know the Russians have done from your observations, not what they have told us they have done—what you have observed they have done. Would you be in a position to state whether or not they had developed themselves in these other areas that you have just talked about, besides the area of getting a thrust and push?

Dr. DRYDEN. I think the best evidence on the open record are the perigee altitudes of the three sputniks according to the final computations made in this country: 142 miles, 140 miles, 135 miles. Perhaps I should read the Explorer's perigee altitude to give some comparison.

The Explorer I, 224; Explorer III, 121; Explorer IV, 163. I think it is evident that the guidance equipment in the sputniks is more accurate than that in Explorer. Let me remind you, however, that we could not put in the Explorer our ballistic guidance system because it weighed too much for Explorer. I don't want my statement to be interpreted as meaning that the United States does not have equally good guidance. But there is the positive evidence that the Russians do have good guidance. They were able to put up three satellites, as you can see, with a spread of 7 miles in height.

Mr. KING. Is there any reasonable possibility that the Russians might have been very lucky, in other words, that lacking the advanced state of the art that you are talking——

Dr. DRYDEN. In two tries they might, but in three, no.

Dr. PICKERING. The inclination at the equator is another indication. They come out 65, 65, and 65.3 degrees.

Mr. KING. The author of this article stated among other things, as I recall, that he checked with all responsible Government officials who would have any knowledge or alleged knowledge of Lunik and decided on the basis of his conversations with them that there was no positive evidence that there was a Lunik, or at least that it did what it was alleged to have done. My question is: Did the author of the article contact you, Dr. Pickering, or anyone associated with your staff, to your knowledge?

Dr. PICKERING. I believe he talked with one of the men who is on the tracking dish after publication.

Mr. KING. Did he before publication?

Dr. PICKERING. Not to anyone in the laboratory.

The CHAIRMAN. How does the data that you have just given us in reference to the Russians, regarding the similarity in each one of their efforts, how does that compare with our own?

Dr. PICKERING. If we look at the Explorers, if we take the experimental evidence of what the Explorer altitudes are, they come out to range from 224 to 121. That is a range of 120 miles in perigee altitude, as compared to the Russians having a range of only 9 miles in perigee altitude. One has to make two assumptions, first we were trying to put all of ours to the same perigee altitude, and the Russians were trying to put all theirs to the same altitude. This does not mean that we don't have better guidance than we used in the Explorers, as Dr. Dryden said.

The CHAIRMAN. We were using our second team?

Dr. PICKERING. Explorer, after all, was in many ways a simple rocket. What was done was to take a first stage booster and to mount a three-stage solid propellant rocket. The three stages were guided only by the fact that they were spinning. You have this spinning rocket, three stages of it fired out——

The CHAIRMAN. No separate guidance system?

Dr. PICKERING. That is right, except for the spinning.

Dr. DRYDEN. The reason is that we did not have the booster capacity to put up the good guidance systems that we now have.

The CHAIRMAN. The Russians have that?

Dr. DRYDEN. Yes.

The CHAIRMAN. My colleagues have suggested this: Are there any American scientists of repute that disagree with your findings in reference to lunik?

Dr. DRYDEN. Not to my knowledge. I think, Mr. Chairman, you are to hear in closed session other branches of the Government who do have additional information to give to you on this subject.

The CHAIRMAN. Here in the presence of the press, your answer on that is "No"?

Dr. DRYDEN. Not to my knowledge.

The CHAIRMAN. To your knowledge there is nobody that disagrees with you about the fact that lunik went up in space and at least passed the moon?

Dr. DRYDEN. I mention again, Mr. Chairman, the publication in the Journal of the Russian Academy of Sciences under date of April 11, 1959. The article at the end says it was submitted on February 25. Both the submission date and the date of publication are such that they could not have had Dr. Van Allen's analysis of his final measurements on our later satellites and space probe. They do give one reference to Dr. Van Allen's report at the fifth assembly in Moscow which Dr. Newell has referred to. The article gives charts and diagrams of their measurements with four instruments, two Geiger counters and two scintillation counters out to a distance of about 60,000 miles. They talk about it as the early analysis of their data.

This is in quite a lot of detail and agrees in its general features with the results that we have obtained with our instruments.

The CHAIRMAN. As an independent thinker you reached certain conclusions and later on you obtained that article and found it fits in with your own conclusions?

Dr. DRYDEN. Yes. These were published before our own measurements could have possibly been available to the Russians.

Mr. SISK. I would like to get back to what Dr. Newell was discussing a moment ago when he was talking about reading some excerpts from this paper. I have one particular paragraph that I had noted here, Dr. Newell. This happens to be by J. W. Townsend, Jr. In his concluding paragraph he indicates that at the meeting U.S. data in the field commanded more respect than U.S.S.R. data. It was given in more detail and covered broader effort.

He says:

> At the next intimate gathering of this type, the Soviet results will be on a par with ours and unless conditions change, the Russians will be ahead.

I am interested in what your impression is and that of the people who accompanied you as to the progress the Russians are making in the field of instruments, vehicles, guidance, and all the other things. To me it has significance here.

Dr. NEWELL. I would like, in answer to this question, to read a few more excerpts from this paper to illustrate what the Russians did produce, and then come back to a comment which Mr. Sisk——

Mr. MILLER. I suggest he proceed in his own way.

The CHAIRMAN. Proceed.

Dr. NEWELL. On page 3, under the heading "Ionosphere and Positive Ions," I would like to point out:

> The Soviet papers on the distribution of electron density in the ionosphere and on the nature and number of positive ions were impressive. The Soviet rocket and satellite experiments have apparently provided the first quantitative data on the profile of the ionosphere above the maximum of the F_2 region.

On page 3 on the right-hand column, Mr. Townsend points out that a mass spectrometer was flown on Sputnik III and a little later in that paragraph he quotes the results:

The results given from the first several orbits show that the principal positive ion is that of atomic oxygen, mass 16, in the region from 250 to 950 kilometers.

And he mentioned other things. I would like to point out these 950 kilometers height was above anything we had gotten at that time.

Then over on the next page Townsend points out:

The Soviets have flown a large number of their so-called meteorological rockets during the IGY from launching sites in the Antarctic (Mirny), in the Arctic (Franz Josef Land), and aboard ship at many latitudes. Results on the temperature distribution up to an altitude of 60 kilometers were presented.

In comment I would like to point out we are still working on the development of a meteorological rocket.

Mr. TEAGUE. How do you know that statement is true?

Dr. NEWELL. We have the papers. We have the results from their papers. I brought along the package to show you how many papers we got from this conference. In my estimation it is not possible for them to invent the data that they presented.

Then he compares these results with ours. And later on that same page 4 he says:

The data show the same overall temperature distribution with altitude as is shown in U.S. data. However, for the first time, complete curves were presented, apparently from a large number of firings in a complete program, showing the seasonal variation of temperature at each of six 5-kilometer intervals between 25 and 50 kilometers.

He continues in this vein. The point I am trying to bring out is that the Soviets are dealing with the same types of problems that we are dealing with. They are solving them as well as we are solving them. But in addition they seem to be putting more attention on this effort than we are.

As an example, in Mr. Townsend's group here in the United States he had at that time four people working on this upper atmosphere pressure-density material. In conversations with their Russian counterparts, they found that they had 10 times as many people on that one subject.

In other words, 40 people working in this single area. This then is the basis of Mr. Townsend's judgment that unless conditions change, the Russians in the future will be ahead of us. The significant point is not that we and they are about equal now, but that the Russians are moving ahead faster, at least in our estimation.

Dr. DRYDEN. May I insert a footnote? Mr. Townsend has more people now. This was written before the creation of NASA, when he was at the Naval Research Laboratory. We think conditions are changing.

Mr. FULTON. I think you should state the specific fields you are speaking of in comparing the Russians to the United States and not just leave hanging in the air a general statement that unless we go faster they will be ahead in the future. I think you should pin it down into a point in time, and likewise give the circumference of the fields you are speaking of.

I don't think you are qualified, I might add, to make a general judgment from the knowledge you have. My reason for saying that is that on page 2 of the IGY pamphlet which you have in your hand, Dr. Newell says this:

> Policies for the operation of the world data centers and the exchange of IGY data developed rather smoothly except in the area of rockets and satellites. Difficulties arose because of the U.S.S.R.'s refusal (1) to provide orbital elements for Soviet satellites during the course of the satellites' lifetimes; (2) to provide precision radio tracking data for satellites; (3) to agree to an automatic dispatch of basic data to the world data centers. On the last point the U.S.S.R. provided no guarantee whatsoever that the rest of the world will ever see any of the desired data; the Soviets say that they intend to negotiate each request for information with the requester. When it was pointed out that this procedure would not provide a means for scientists to know what data existed, the Soviets had no comment. As acting reporter for rockets and satellites, I felt compelled to state in my report to the full assembly on the last day that the working group for rockets and satellites had failed to achieve a satisfactory solution of the data exchange and world data centers problem.

I would like you to correlate those two statements for me. It appears you have enough information, by the statement of Dr. Newell as of January 1959, but the contrary would appear to be the case. If we have sufficient data without the Russians cooperating, it is ridiculous for us to be demanding their cooperation. If we don't get sufficient data without Russia's cooperation, and they are not cooperating, I don't think we should give a valid overall judgment on who is ahead or who is behind. Would you please comment?

Dr. NEWELL. The area refers to the area of rocket and satellite research in outer space——

Mr. FULTON. Aren't you talking about that here?

Dr. NEWELL. This delimits the comments I am making with regard to their abilities. As to the point you have just brought up, the key phrase here is that the Russians refused to agree to an automatic dispatch of basic data to the world data centers. This is an important point, because it means that they then propose to choose and select both the data they will dispatch to the world data centers, and the time at which they will dispatch this data.

It is true that in the Moscow Conference they had a large number of papers, as I indicated here. But the papers had a certain character to them, which is distressing to the scientists, namely, that they simply reported basic research results without, also, in many cases adding the fundamental data that one needs in order to judge whether or not their conclusions are correct.

Mr. FULTON. I hope then you will limit the broad judgment you gave just previously.

Dr. NEWELL. This was the point I was about to come to. I have carefully pointed out in my answer to Mr. Sisk that the basis for our judgment was that the Soviets were dealing with the same sorts of problems. They were faced with the same problems in research. This is the key. When you ask what a scientist is doing, how well he is doing, the answer comes out in the sorts of problems he is attempting to attack. This is the strongest point in our judgment. They are dealing with the same sorts of problems we are. They are up with us.

Mr. FULTON. In what fields?

Dr. NEWELL. In this area of upper atmosphere and space research with rockets and satellites.

Mr. FULTON. You aren't including the ICBM's or——

Mr. SISK. Will the gentleman yield?

Mr. FULTON. Let him answer that.

Dr. NEWELL. No; I am not.

Mr. SISK. I was restricting it strictly to what this paper had in mind. That is what we had under discussion. I did not mean by my question to indicate anything with reference to missiles, or defense or security. I was strictly talking from the standpoint of acquiring data of upper atmospheric conditions and all the multitude of things that go with this type of work.

Mr. FULTON. May I finish on the quotation I had?

If we are able to come up with the data on the orbital elements to the point of saying that the declination from the Equator is at 65 degrees, 65 degrees and 65.3 degrees, without the Russians giving it to us, why do we need to comment on the fact that the Russians will not cooperate, as you have done in this particular article?

Dr. NEWELL. The orbital elements for a satellite trajectory are six in number. In order to pin down exactly where the satellite will be at a given time you need to know all six of those elements. The two that they gave are the very simplest, the ones that one can get by rough observation.

Mr. FULTON. Is that enough for you to say that they have good guidance and control equipment that would be equal to the United States in that particular field?

Dr. NEWELL. Much of the data they presented indicated they did have very precise equipment.

Mr. FULTON. Equal to that of the United States in this particular field?

Dr. NEWELL. This was our feeling. I must emphasize though it was a subjective feeling at the time.

Mr. FULTON. Your real criticism of the Russians' failure to cooperate is one rather of time and degree and not that they really just didn't cooperate at all? They might have held back a little bit, so in all justice to the Russians we would have to say that they didn't come up to your expectations, because you are able to make valid judgments from the scientific information they have given.

Dr. NEWELL. The criticism that we make is a criticism which most of the rest of the attendees at the IGY Conference also made, that in order to have an open and really effective interchange of data, this must be automatic, and agreed to on the basis of all the countries.

The CHAIRMAN. As I understand—I want to see if I understand you correctly—you say that one of the most important means you have of estimating the Russians' capability in this field is the fact that they are handling the same problems that we have.

For instance, is this correct: that the Russians would not have a guidance problem unless they were able to get the missile off the ground, and then they would not have problems of man-in-space unless they were able to get a missile up in orbit, and then undertake the problem of the man in space, is that correct?

Dr. NEWELL. These are analogous cases. I am applying my remarks to upper air and space research. But the analogies you give are similar; yes.

Mr. KARTH. I was interested in Dr. Dryden's answer to the series of questions that Mr. King had. Let me ask you this: If we had the proper booster power and used our best guidance system, so that we could get the proper weight in space, could we do better than the 9-mile perigee difference that the Russians apparently did, or could we do equally well?

Dr. DRYDEN. I think we ought to answer that one in closed session.

Mr. KARTH. One other question: Dr. Newell, Mr. Mallan said he went into a technical bookstore and, in his opinion this was a bookstore where he could feel he was getting the proper information insofar as the technical advancement was concerned, that he saw diagrams and graphs in books, I believe, that were dated 1958 or at least very late in issue, and from those diagrams and from what he read he decided their electronics equipment and their receiving equipment were very primitive, or he said at least comparable to ours or what ours was in 1950 or 1952. Does that make sense to you, and would you agree with that statement or not, insofar as receiving and electronic equipment?

Dr. NEWELL. I can't base my remark on the particular book that he has seen. The electronic equipment that we saw in connection with the Sputnik III exhibit was well constructed. Some of them used transistor construction.

Mr. KARTH. Did you see any of the receiving equipment that was comparable to ours?

Dr. NEWELL. I did not see any receivers. I saw the transmitters, the gages, and so on, in Sputnik III. I do not recall seeing any receivers.

Mr. KARTH. If you have enough booster power to reach escape velocity wouldn't it orbit the sun?

Dr. NEWELL. Yes.

Mr. KARTH. If they had the booster power and did have the escape velocity it is conceivable to think it did orbit the sun?

Dr. NEWELL. That is correct.

Mr. FULTON. When you have a freely moving body in the planetary system we are still not able to explain comet reactions. It may orbit Jupiter or something else.

Dr. PICKERING. If one starts out from the vicinity of the Earth with a velocity outside the Earth's gravitational field, which is relatively small, by "relatively small" I now mean a few thousand miles an hour, then you must inevitably go into an orbit around the Sun which is not too different than the Earth's orbit.

Mr. FULTON. It could orbit Jupiter.

Dr. PICKERING. It would have to have higher velocity to go to Jupiter and be trapped. In order to do this you would have to leave the Earth's gravitational field with a quite high velocity.

Mr. FULTON. Then don't all comets orbit the Sun? Don't they just have a perihelion and go off into distance?

Dr. PICKERING. Many of them do have a very extended elliptical orbit. Haley's comet takes 75 years——

Mr. FULTON. Not all of them.

Dr. PICKERING. No; some come from outer space and whip around the Sun and go back to outer space, where outer space is now outside the solar system. It takes high speed to do that.

Mr. FULTON. There are objects in the planetary system that, once they get beyond the gravitational field of a body, do come in and go out.

Dr. PICKERING. I think we might look at it this way: Take the Earth, and we put an object around the Earth into a satellite orbit. This will take a certain speed to make just a circular orbit. Suppose one throws it with higher speed. The circle becomes an ellipse and as you increase the speed the ellipse gets more elongated and finally opens up and off you go. It is the same sort of situation here.

Mr. FULTON. I didn't want it in the record that anybody that had escape velocity from the Earth in all cases orbited the Sun.

Dr. NEWELL. I am sorry, Mr. Fulton, but you are wrong there. Anybody that——

Mr. FULTON. I think I am right. I could shoot something at Jupiter and Pluto and keep it orbiting there.

Dr. NEWELL. It is still orbiting the Sun. You have a variety of different types of orbits. You have closed orbits, open orbits, orbits that are modified. Technically the term "orbit" refers to the path of one body around another gravitating body. We have in the case of the normal concept of an orbit ellipses and circles. The comets to which you refer sometimes follow parabolic or hyperbolic orbits. There are other orbits that go so close to big bodies like Jupiter that the path is distorted from any one of the ellipses.

The CHAIRMAN. I think I ought to refer that dispute to a subcommittee——

Dr. DRYDEN. I think it might help to say that the Moon also orbits the Sun. The path in space is a path like this: A body which orbited Jupiter could do the same thing.

Mr. FULTON. Obviously we were all talking about a free orbit of an artificial satellite around the Sun.

The CHAIRMAN. Coming back down to Earth, as we have been quite far out, in reference to scientists generally, don't we have a number of our own people who have been over to Russia and interviewed Russian scientists and stayed there for some time and talked with them?

Dr. DRYDEN. I would suggest that you get Dr. John Tukey from Princeton to come down and talk to the committee sometime.

The CHAIRMAN. A representative from Tulane in my own State came in and said he had been over there several months working with the Russian scientists. Don't we have a number of them that have done that and can give this committee a very factual account of the capabilities of the Russian scientific system over there? We shouldn't have any trouble getting the data we need in that respect; should we?

Any further questions?

Mr. MITCHELL. Yesterday, as I recall, a statement was made to the effect that Sputnik III carried the largest payload. We only have their word that they in fact launched such a heavy payload; is that correct?

Dr. PICKERING. There is obviously no way in which we can tell exactly what the payload is, except what the Russians tell us. However, one can draw some conclusions as to the approximate weight of the payload. For example, knowing the path of the sputniks there will be a certain amount of drag exerted by the residual atmosphere at those altitudes. This drag will gradually bring them down so after a few months they return to earth and burn up.

The rate at which this occurs is determined by the area of the sputnik in relation to its mass and to its weight.

In other words, if you have a very heavy, compact body there won't be much drag. If it is a very spread out body the drag will be more effective for the same weight. Then one can say that by taking photographs through telescopes of the sputniks we can estimate the size of the rockets which are in orbit, for example. I might point out in a general way they are about the size of the Atlas, the rockets in orbit. So having this size, you can then say, knowing the size and knowing how fast it comes down, you can then estimate the weight. It is not a precise estimate, but it gives you answers which are consistent with what the Russians stated.

Mr. MITCHELL. Were your observations consistent with the Russian announcement of the payload?

Dr. PICKERING. Yes, sir.

Mr. FULTON. I want to question you on one other aspect of orbiting the sun. Obviously every satellite that orbits the earth is also orbiting the sun; is it not? But you didn't mean that when you said this satellite that the Russians shot out and went beyond the moon orbited the earth or any other body and orbited the sun. You meant it went into a free orbit of its own; did you not?

Dr. PICKERING. It went into an elliptical orbit around the sun.

Mr. FULTON. The question comes up as to the velocity that that particular missile or vehicle was going at the time it cleared the earth's gravitational field. Do you know anything about the velocity? If you don't, it may not have had enough velocity to orbit the sun in any sort of elliptical orbit and Kepler's law might have operated for a time and it might have dropped into the sun.

Dr. PICKERING. It would not drop right into the sun. If it goes away from the earth at all, let's say it has enough speed to get out of the earth's gravitational field and stays out. Then since the earth is traveling around the sun at a speed of roughly 100,000 feet a second, 18½ miles a second, this means that a body——

Mr. FULTON. Don't combine with each other against me.

Dr. PICKERING. I was checking my number. What I am saying is that a body which is just outside the earth is likewise traveling around the sun at this speed of 100,000 feet a second because that is the speed that the earth is traveling at.

Mr. FULTON. You don't know with what velocity it left the earth's gravitational field.

Dr. PICKERING. Suppose it left with zero speed. Then it would still have this speed of 100,000 feet a second around the sun because that is what the earth has around the sun. Therefore it would go around the sun in an orbit that would be very close to the earth's orbit.

If it leaves the earth's gravitational field with a speed which is a few thousand feet a second, then it will go into an ellipse around the sun which is fairly close to the earth's ellipse. If I remember rightly the lunik figures were something like 140 million miles out as against 90 million miles closest approach to the sun. The Pioneer IV figures were something like 120 million miles against 90 million miles. They are the same order of magnitude. In both cases the object left the earth's gravitational field with a speed of only a few thousand feet a second.

Mr. FULTON. You would say, then, it would not have the tendency, because of the sun's gravity, to drop——

Dr. PICKERING. Because it has the speed of the earth. If you wanted it to drop into the sun you would have to shoot it backward along the earth's speed and it would have to have a speed of 100,000 feet a second. If you could leave the earth's gravitational field and still have a speed of 100,000 feet a second and shoot just backwards from the direction the earth was traveling, then you would fall straight into the sun.

It turns out to be one of the most difficult problems to do. From a scientific point of view you would like to do this, to put an instrument into the sun. It is a tough problem.

Mr. MILLER. I appreciate and love to hear these gentlemen talk; but we didn't come here to study astronomy and get abstract lectures. We have some other things to do. I think we ought to go into executive session now and get on with our regular work.

The CHAIRMAN. Is there any more testimony that can be given in open session? If not, then we agreed to hear these gentlemen in executive session. The committee will accordingly go into executive session.

(Whereupon, the committee proceeded in executive session.)

Next 7 Page(s) In Document Exempt

SOVIET SPACE TECHNOLOGY

WEDNESDAY, MAY 13, 1959

HOUSE OF REPRESENTATIVES,
COMMITTEE ON SCIENCE AND ASTRONAUTICS,
Washington, D.C.

The committee met at 10:15 a.m., in room B-214, New House Office Building, Washington, D.C., Hon. Ken Hechler, presiding.

Mr. HECHLER. The committee will come to order.

The first witness this morning is Dr. S. Fred Singer, associate professor of physics, University of Maryland.

Dr. Singer, if you have a prepared statement, you may proceed.

STATEMENT OF DR. S. FRED SINGER, ASSOCIATE PROFESSOR OF PHYSICS, UNIVERSITY OF MARYLAND

Dr. SINGER. Yes, sir, I have a very short statement.

Gentlemen, the purpose of these hearings, as I understand it, is to gain a better assessment of Soviet advances in space science and space technology and of what we must do in order to surpass them. It is particularly important therefore to be able to assess Soviet claims. For example, Mr. Lloyd Mallan, in an article in True magazine, has put forward the idea that the Soviets never launched a lunik and that the whole thing is a hoax. I do not subscribe to this view simply because there is more evidence to the contrary.

But while we must be, of course, careful not to underestimate Soviet achievements, we should not go to the other extreme of being overly impressed with them. I have made a careful study of some of the aspects of the Soviet lunik, and I would like to give you very briefly some of the conclusions.

No. 1. I find that there is very adequate evidence that the rocket was in fact launched;

No. 2. I find very adequate evidence that at least some of the scientific instrumentation in the rocket worked;

No. 3. I find evidence that the Soviets intended the rocket to hit the moon but missed it.

To account for some of the astounding conclusions of Mr. Mallan we must remember that space technology in the Soviet Union is carried out under a deep cover of secrecy. None of the scientists who actually work on rockets have ever been publicly identified. Instead the Soviets have issued misleading statements concerning people who are supposed to be responsible for sputnik. But on the contrary, with what amounts to absolute stupidity, the Soviets have not made use of some of their ablest scientists in their space programs. We know this because we can read the publications of these scientists and find that they rely on Western experiments in order to draw their theoretical conclusions.

The Soviets will, in all likelihood, use their large rockets for heavier payloads and they should also be in a position to launch one or more men into orbit in the very near future. While this feat will have a considerable propaganda effect, I don't think it is going to affect the military balance between East and West.

On the contrary, I have absolutely no doubt that we are going to surpass the Russians in space technology as we are already doing in some more restricted fields, such as instrumentation, electronics, tracking, and so forth. But we must be sure not to relax our efforts. Right now there are excellent scientists involved in our programs and they have a good amount of support. This must be continued. We must also remember that we in the United States do not have a monopoly on scientific brains and should draw on Europe and Latin America to help us in advancing into the space age.

Thank you.

Mr. HECHLER. Thank you, Dr. Singer.

Do members of the committee have any questions?

Mr. FULTON. Could I welcome Dr. Singer, and say it has been my pleasure to have been associated with him in the past on this work? I would like to compliment him on his success in his studies in the space field.

Doctor, when you said that the Russians would be able to launch men into space in the very near future, could you be more exact as to what you mean by that term, and whether it is in orbit or merely an IRBM shot?

Dr. SINGER. I do mean in orbit. Let me be very precise about it. I think the really difficult technical problem about putting a man in orbit is to make sure that the rocket that does the job is reliable. I don't know of any way of making sure that you have a reliable rocket until you fire it, say, a hundred times, and it worked 100 times or 99 times. Only then do you know that it is reliable. This takes a long time. The Russians have been firing large rockets for some time. They themselves claim that they have never missed. That is their claim. Therefore, I would ask them the following question: Why haven't you put up a man? You have had 2 years, which is long enough, I think, to construct the capsule.

We in our program here have had proposals for manned satellites. I know of proposals as long as a year and a half ago, seriously, well worked out proposals, which only required a reliable rocket in order to carry them into reality.

Judging from our own time scale, which is Project Mercury, about 18 months, I would say that the Russians should be able to put a man into orbit in the very near future.

Mr. FULTON. What would the words "very near future" be in your lexicon? What does that mean?

Dr. SINGER. I should think in months, measured in months.

Mr. FULTON. That would mean before or after our 18-month period for Mercury?

Dr. SINGER. I believe before.

Mr. FULTON. How much in advance do you think they might be, in your judgment, over us on a Mercury-type project? How much before do you think they might be?

Dr. SINGER. If we assume that it takes them as long to construct the capsule and work out the technical problems associated with use, which is about 18 months, then by this simple calculation, they should have launched one by now.

Mr. FULTON. Of course, I could answer for the Russians when you say it takes a big enough and dependable enough rocket to get them up. They say they now have such a rocket. The Russians might say, yes, but how do you get him down? You still have to get the man down.

Dr. SINGER. Yes, that is quite true, and, of course, there are many technical problems associated with putting a man in orbit and recovering him successfully, but the greatest problem by far in expert opinion seems to be to make sure that the launching rocket is reliable.

Mr. FULTON. Do you believe that the Russians have the ability to recover a descending orbiting vehicle?

Dr. SINGER. Well, we don't know, of course, but I think we should assume that their capabilities in this area are roughly the same as ours. This would be the most prudent assumption.

Mr. FULTON. Would you comment on our type of approach to the Mercury program specifically in regard to the space vehicle? As I understand it, you have written articles and have been interested for some time in space vehicles. Would you comment on whether you believe our Mercury approach is adequate and is it the best?

Dr. SINGER. This is a very difficult problem to decide. There are a number of ways in which one can bring back a capsule, and this, of course, reflects on the construction of the capsule. One can build a capsule which uses a heat shield. One can build a capsule which uses an ablation shield. One can build a capsule which uses what amounts to a large parachute which radiates the heat away. The problem in all cases is to somehow keep the man inside the capsule from being burned up.

I think the present approach probably represents something which is close to optimum. This is why it has been chosen. As always, there are technical opinions involved and everybody doesn't see eye to eye. I don't have any strong technical objections to the present program, but I would say this: I would be much happier personally if we were not to put all of our eggs into one basket.

I would like in this particular field of capsule construction to see more than one program. I would like to see at least one competing program using a different technical approach, just for the same reason that we have had different technical approaches in other enterprises.

I only have to remind you of the fact that if we hadn't had the Jupiter rocket to back up the Vanguard we probably would have been much, much later putting a satellite up into space.

Mr. FULTON. I am interested to hear you say that, because I have objected myself to cutting it to 1 company when 36 companies had shown interest and had various avenues of approach. For example, Avco had a different method and was summarily cut out, so all their research has gone for naught at this time.

Would you recommend there be a broader base of competition and selectivity so we narrow it down possibly from 36 companies to 25, to 20, to 15, to 12, and until you may have them working on combined programs that are competitive and, nevertheless, having different avenues of approach, so that we get a better coverage of the scientific field?

Dr. SINGER. I believe the only proper comment for me to make is to say as a general comment that I would be happier by far if there were more than one technical approach to the capsule problem. I would like to see competing approaches being pursued, and the reason for this duplication is the following: It is not really wasteful. By far the largest amount of money is going to be spent on the testing of the booster rocket, which is going to be the same for all of the programs. The capsule represents a rather small expenditure in money, but it is, of course, the vital part of the program.

Mr. FULTON. How far do you think the lunik got, in your view, when you judge what signals have been received with tracking? What has been the result of that shot of the Soviets?

Dr. SINGER. Here I must say the following: The only information which has been available to me is the published Russian information. Whatever information has been picked up in the United States about the orbit of lunik, or whatever information has been received from lunik by way of tracking equipment, has never been released.

Mr. FULTON. Well, when the signal stopped, why do you think it stopped?

Dr. SINGER. There could be several reasons. No. 1, there might be a real failure of the equipment. This is possible, but may not have happened. For example, a meteor could have hit the transmitter and knocked it out. No. 2, the batteries could wear out, and the signal then gets very weak. No. 3, the distance gets greater and therefore the signal becomes weaker. Any one of these three causes could have been responsible for a signal disappearing.

Mr. FULTON. There is no reason that we couldn't have received that signal further?

Dr. SINGER. Well, you can decide between the possibilities by watching the signal before it disappears. If it has been getting progressively weaker, then I would rule out cause No. 1, but I should say again that I have not been able to get any information about the signals from the Russian lunik. They have been rather closely kept.

I read Mr. Mallan's articles before I came here, and he made a remark in there that he questioned a general concerning any signals from lunik, and all he got from the general was an enigmatic smile. I think if they had actually picked up the signals, it would have been better to have told Mr. Mallan that and he wouldn't have published his article in this form.

Mr. FULTON. You see, I am trying to get your judgment on what kind of a shot this was and how far it went and how much we are sure of. You see, I am trying to get your estimate now from what you have seen and read.

Dr. SINGER. From the information the Russians have published?

Mr. FULTON. From whatever source. I want your statement.

Dr. SINGER. We can be sure that the rocket went out, I would say, about half the distance to the moon, at least, but that is all we can be sure of.

Mr. FULTON. So that as far as the Russians claiming the first vehicle to orbit the sun, that is just somebody's guess, and likewise any scientist in the United States who says that they were the first to orbit the sun would be using a measure of guesswork, too.

Dr. SINGER. This is their claim, and if you believe their claim, that is fine.

Mr. FULTON. Do you believe that there is adequate evidence to say the Russians are the first through the lunik to have orbited the sun? I don't. I have been trying to prove that here in the last few days.

Dr. SINGER. I can only believe this when I see the actual data which show the tracking information received from the lunik all the way out past the moon.

Mr. FULTON. So that at the present time there is no reason for the U.S. scientists to be so ready to admit that the Russians have in orbit around the sun a vehicle of 700-pound size?

Dr. SINGER. I would cautiously agree there, except for the following fact: I think the Russians would not tell a lie if they thought it could be found out. Now, the rocket——

Mr. FULTON. I disagree with you on that.

Dr. SINGER. If they thought it could be found out, if they could be uncovered.

Mr. FULTON. Haven't they claimed many firsts of science that, of course, they have already admitted to be firsts in other countries of the West?

Dr. SINGER. Well, they would look very silly if we in fact did track the rocket and showed it coming back, because if it did not escape from the earth, it must, of course, come back. So they are taking a real risk here in making a claim which could then not be substantiated later on.

Mr. FULTON. But so far we don't know what kind of ellipse that rocket went into, and even on our own recoveries we were uncertain about the fact that the vehicle had at least descended. Isn't there some uncertainty?

Dr. SINGER. Yes, I think there is some uncertainty. I think the Russians themselves are very uncertain about the orbit of the rocket. They probably originally thought it was going to hit the moon. They called it the lunik. They also announced in the original release one of the purposes of the rocket was to measure the radioactivity of the moon's surface and the moon's magnetic field. You can only do this if you come very, very close to the moon. In fact, if you want to measure that you have to actually hit it. This was in the initial release. Then a little later on the name of the rocket was changed from lunik to mechta.

I talked to one of the local experts and asked him what it meant, and he said mechta, why, that means a dream, sort of like an unattainable ideal, and soon thereafter, of course, they announced that it had become the 10th planet of the solar system, which I think is taking a lot of license, because in addition to nine planets there are millions and millions of asteroids going around the sun. So even if it had gone into orbit around the sun, it would not be No. 10, but it would probably be No. 2,500,000 and so on.

I might add one more point. About 2 years ago, Prof. Fritz Zwicky, of California Tech, in an Aerobee rocket released a shaped charge. The pellets of the shaped charge must have gone into orbit around the sun. If you want to give the first to a manmade object to orbit around the sun, I would give it to the shaped charge in the Aerobee rocked fired at White Sands.

Mr. FULTON. I am glad to hear you say that, because I think some of the scientists in the United States have been premature in conceding that Russia is the first nation to have a vehicle in orbit around the sun.

Thank you. That is all.

Mr. MITCHELL. Dr. Singer, I think in response to Mr. Fulton's question, you gave some of your reasons for believing it was the intent of the Soviets to strike the moon. Will you elaborate further along that line? That was one of the three points made in the beginning of your statement. Why do you feel it was their purpose to strike the moon?

Dr. SINGER. When they made their announcement, they said that they intended to come close to the moon. Now, I don't believe anything they say as such. I would have been much happier if they had made this statement before they launched the rocket, if they had announced this distance which they claimed before they launched the rocket. This they didn't do, of course, but they simply observed the orbit, and after they decided it was going to miss the moon they calculated by how much, and then released a number which, as far as I know, would be very difficult to check. Perhaps it is being checked by tracking observations that have been taken by various military groups around the country.

The reason, of course, why I thought they intended to hit the moon is based entirely on the scientific purposes which they stated for the rocket. They stated they had equipment aboard to measure the surface radioactivity of the moon and the moon's magnetic field.

Now, the radioactivity of the moon decreases as the inverse square of the distance from the center of the moon. The magnetic field, if it exists at all, falls off much faster, all of which means that if you really want to make these measurements, as they claim they did, you would have to come very much closer to the moon than they did, in fact. They came within about 3½ lunar radii, they say. Instead of 4,500 miles, they would have had to come within a hundred miles to make these measurements, which means that for all practical purposes they would have had to hit the moon.

Mr. MITCHELL. Dr. Singer, you stated, I believe, in your testimony that your belief as to the launching of the lunik is based solely upon data that has been released by the Russians; is that correct?

Dr. SINGER. That is correct.

Mr. MITCHELL. And you made a statement, as I understood it, that from this data you think they substantiate the fact that the firing went at least halfway to the moon?

Dr. SINGER. Yes.

Mr. MITCHELL. What do you base that statement on?

Dr. SINGER. An article was published in the proceedings of the Russian Academy of Sciences in April, last month, which was called "A study of the terrestrial corpuscular radiation and cosmic rays during flight of the cosmic rocket." Notice they now call it a cosmic rocket and not lunik anymore.

They plot the trajectory of the rocket as far out as 120,000 kilometers, which is roughly 20 earth radii, or a distance of about one-third the distance to the moon.

Mr. MITCHELL. Doctor, have you been to Russia and had a chance to observe any of their equipment, their rockets?

Dr. SINGER. No, sir; I have not. I intend to go this summer on invitation of the Academy of Sciences for a scientific meeting, and I hope there will be some opportunity to see some of the equipment myself.

Mr. MITCHELL. Have you talked with others who have had that opportunity?

Dr. SINGER. Yes; I have.

Mr. MITCHELL. I am generally discussing the article by Mr. Mallan and the conclusion that the Soviets do not have an operational ICBM, that in that area they are many years behind. Do you feel that you can base such an opinion on facts, or that you have sufficient information to give such an opinion?

Dr. SINGER. I have no privileged information, but I can make some general statements. I don't think Mr. Mallan is justified in drawing his conclusion from the facts that he quotes. However, his conclusion may still be correct. It depends on what one means by operational. The Russian claim is that they have an operational ICBM. I would not take that very seriously because it is just a claim. They have not demonstrated it in any way. They have not yet demonstrated that they have the ability to guide accurately. They have not demonstrated that they have the ability to recover a warhead, bring it back intact, and thereafter make sure that the bomb inside the warhead will not be damaged.

We should assume, of course, just as a matter of being prudent about it, that they do have this capability, but we cannot be sure of it. But to have an operational ICBM involves very much more. You have to have a ground complex, ground support, to enable you to launch hundreds of these all at the same time.

As we know, the strategy of warfare using ICBM's is not to launch a single one, but you must launch hundreds of them at the same time against all conceivable targets to wipe out, if you can, all centers of retaliation.

The Russians have a very large ICBM, one which uses approximately 500,000 pounds of propellants. If they wanted to launch, let's say, just 200 of these in a single attack—and this is what I mean by an operational ICBM, operationally used—they will have to supply a hundred million pounds of propellant, fuel them up all at the same time, and get them ready and get them launched within a matter of minutes, and this is a big operation. I am not saying that they could not do it, but it is something that would take a good part of their national product.

Mr. MITCHELL. That is all.

Mr. MOELLER. Dr. Singer, I gather from the comments you made before that you are not quite certain of the guidance ability of the Russian devices. I think we had heard testimony a number of times here that Russia does have thrust but it doesn't have guidance ability for its devices. But recently we heard on the basis of the performance of the sputniks that evidently they have good guidance. Does this agree with what you have gathered?

Dr. SINGER. If we knew the orbit that they intended to take, then we would know, of course, how good their guidance is. They have never announced the orbits in advance, and therefore it is very hard to judge how close they got to their goal. However, assuming that they did mean to hit the moon, then lunik gives us somewhat of a gage to judge how good their guidance is. I would say from this that their guidance is—I would add that this, of course, as you know is a minimum estimate—their guidance could be much better than that. We will never know. But we can certainly say that the guidance they have is marginal as far as ICBM's are concerned.

As you know, the guidance for ICBM's doesn't have to be awfully good because the damage radius of the bomb is so large. So you don't have to hit with a fraction of a mile. It is good enough. I mean, you can accomplish your purpose by hitting within a few miles.

Mr. MOELLER. Then you would be willing to say you don't feel that they have a good guidance system?

Dr. SINGER. I believe you have expressed it correctly; yes, sir.

Mr. FULTON. Will the gentleman from Ohio yield?

Mr. MOELLER. Yes.

Mr. FULTON. When you hear that the sputniks had a 63° and 63½° inclination from the Equator on their orbit, does that not indicate to you a pretty good control over the orbit through proper guidance procedure and equipment?

Dr. SINGER. The sputniks would indicate that they have the ability to control to within about half a degree. Whether you consider that good or not depends on the purpose. It is good enough for a sputnik. It is not quite good enough for an ICBM if you want real accuracy.

Mr. FULTON. Is that as good as we have on guidance and control of our satellites?

Dr. SINGER. I think it is. I think, however, that our guidance systems are being improved and will end up to be much better than that.

Again, to be prudent, we should assume they are also improving their guidance systems.

I might add one other remark here. We don't know this for sure, of course, but I have read that they adjust the final velocity of the vehicle by radio guidance. In other words, they fire the rocket, they take a quick look and see how well it came up, and then they make some adjustments by simply controlling a little rocket. There is a little rocket still mounted on the main rocket, which acts as a vernier adjustment and is controlled from the ground by means of radio guidance. You can get it pretty close to what you want it to be.

This type of radio guidance, of course, is fine for space vehicles, where you have a lot of time, but it is not very good for an operational ICBM. Therefore, our aim in all of our designs has been to perfect an internal guidance system, an internal guidance system which, in fact, will tell the rocket where to go and which doesn't require any external corrections.

The thought among missile experts is, as far as initial guidance is concerned, we may be somewhat more advanced over the Russian developments.

Mr. MOELLER. May I ask another question?

Mr. HECHLER. Certainly.

Mr. MOELLER. This may have been asked. I am curious to know, Doctor, will you tell us something about your background? Have you already stated that here this morning? If you don't mind, repeat it.

Dr. SINGER. Yes, I am professor of physics at the University of Maryland. I have been there about 6 years. Before that I served 3 years with the naval attaché in Europe. My job was to investigate the development of scientific programs in various countries in Western Europe, to write reports on it, and generally to inform American scientists of what was going on in Western Europe, north Africa, and Scandinavia and so on. So I have had a good deal of experience in assessing scientific programs in foreign countries, and I have traveled a lot.

Mr. MITCHELL. Will the gentleman yield?

Mr. MOELLER. Yes, of course.

Mr. MITCHELL. Will you tell the committee, Doctor, your background insofar as our satellite program is concerned?

Dr. SINGER. Oh, I see. My connection with rockets and outer space programs started soon after I got out of Ohio State. I got an electrical engineering degree at Ohio and got a masters and Ph. D. at Princeton.

In 1946, I joined the applied physics laboratory at Johns Hopkins University. In 1946, we got the first V-2 rockets from Germany, and I was in the first group that instrumented captured V-2 rockets for upper atmosphere research. We measured cosmic rays and emission from the sun in these rockets.

Later on we developed the Aerobee rocket and fired the first units, also at Johns Hopkins University.

Then during my stay in England I became interested in the idea of doing these upper atmosphere experiments in small satellites and propounded the idea of a small-instrumented satellite. At that time the main thinking in astronautics was toward manned space stations. This wasn't going over too well because people realized this wasn't very feasible. My aim, therefore, was to look for a means of making astronautics feasible, but doing it in a practical way, and I thought the application of satellites to upper atmosphere research would be the most useful one, and this has been my connection with satellite programs.

Mr. MITCHELL. What about the feasibility——

Mr. FULTON. Could you yield a minute on his qualifications? Actually the dimensions and the methods of the U.S. satellite programs largely conformed to your advanced predictions and figures; didn't they?

Dr. SINGER. I believe this is approximately correct, except that they have not yet adopted attitude control. I had a system worked out for attitude control which may in fact be going into operation now.

Mr. FULTON. So far it has actually followed your predictions, has it not?

Dr. SINGER. I would say in certain respects; yes. The internal tape recorder and the playback and the solar batteries are certainly coming along.

Mr. FULTON. You were an adviser to the select committee of the House last year; weren't you?

Dr. SINGER. Yes; I was.

Mr. MITCHELL. It is very difficult to bring those things out.

Mr. FULTON. He is a little modest.

Mr. MITCHELL. You stated that you thought it best to proceed with the satellite concept rather than the space-station concept because it was considered not to be feasible at that time.

Dr. SINGER. That is right.

Mr. MITCHELL. With the advancements made now, Doctor, do you consider it feasible?

Dr. SINGER. Yes. As you know, feasibility is nothing absolute. There are things which are not feasible because they violate physical laws. Then there are other things not feasible because they cost too much. Space stations are in the second category. They are perfectly feasible, they don't violate any physical laws, but they cost an awful lot of money and you have to justify them in an economic sense. You have to show people that they are getting something for it. At that time I considered that they were not feasible from that point of view, in the sense that we couldn't persuade people to spend billions of dollars to do it, but I could see spending a few million dollars to build an instrumented satellite, particularly since we are already spending millions of dollars in upper atmosphere research in rockets, and these were going up and coming down again getting only a few minutes' data. A satellite that would stay up for several months would be a very good economical solution to the problem of getting data about the environs of the earth. It has turned out to be that way.

Mr. MITCHELL. Doctor, if I may go to another point. You spent a good deal of time in Europe after the war?

Dr. SINGER. Yes.

Mr. MITCHELL. Did you have an opportunity to meet many of the top Soviet scientists?

Dr. SINGER. Not at that time; no, sir.

Mr. MITCHELL. You have been acquainted with many of them, haven't you?

Dr. SINGER. I am a member of the International Astronautics Federation. Some of the members of the committee were present—I believe Mr. Fulton and the staff members were present at the last meeting held in Amsterdam.

Starting in 1955, I believe, at the Copenhagen meeting, the Russians for the first time sent observers, and in 1956 they formally joined this federation, and Professor Sedov of the Russian Academy of Sciences was made a vice president of the federation.

Mr. MITCHELL. You made a statement earlier that the Russians were not using some of their top men in this field.

Dr. SINGER. That is right.

Mr. MITCHELL. Is that because of your knowledge of who are some of the top men that you make that statement?

Dr. SINGER. Yes, sir. I don't know them personally, but I, of course, read the Russian scientific literature, and I think I can tell a good paper from a mediocre paper. They also publish some very poor papers. I try not to read those.

Mr. MITCHELL. Do you have any opinion as to why they are not using some of their top men? Are they utilizing them, do you think, to their fullest capacity in other fields which they think are equally as valuable?

Dr. SINGER. No, I don't. I think they are not using them because of their great preoccupation with secrecy.

I would like to relate a story, if I may, concerning a personal experience with a very distinguished Russian astronomer, Professor Ogorodnikov, who is a professor of theoretical astronomy in Leningrad. He visited over here last September, at the invitation of our National Academy of Science, and I was his host for the day. I took him around to see the Lincoln and Jefferson Memorials, and I asked him, incidentally, why there were some papers appearing in the Russian literature connected with space subjects, which I thought were not only fairly mediocre but they went over material which had been published in the Western literature many years before, and they didn't quote any references. Of course, I don't know any of these people, so I asked him how come, and he said they are not really academic scientists; they are not used to quoting other people's work. He didn't go into this any more, and I couldn't find out who the people were, but I gathered that they were people working on the Russian war effort. So there must be a sizable body of scientists connected with the Russian war program, probably quite capable scientists. We never hear of them. We hear of them indirectly.

I remember a speech by Khrushchev, a speech he gave in East Germany, not this last visit but the time before, soon after sputnik went up, in which he said something to the effect, "We cannot reveal the names of our glorious scientists who made this glorious sputnik possible. We must protect them." Something to that effect. On the other hand, Pravda did release a long list of Soviet scientists who were cited in connection with sputnik, and you may remember this list. Not one of the scientists on this list, as far as I can tell, and they are well-known scientists, could have had any connection with a rocket program. They are not the sort of people. They are theoretical mathematicians. They are the type of people, if they step into a laboratory all the glassware would shatter. Pravda simply listed their most prominent scientists, of which they have quite a few, and identified them publicly with sputnik, but there is no foundation to this.

Mr. MITCHELL. I might ask one more question. Doctor, you made the statement that in your opinion the Russians would have a man in orbit before we do. Is that primarily based on the fact that they began working on this specific project before we did?

Dr. SINGER. Yes, sir.

Mr. MITCHELL. And would you say that in your judgment they are equally capable of moving along on a time schedule, as we are?

Dr. SINGER. It would seem so.

(Hon. Overton Brooks, chairman, presiding.)

Dr. SINGER. You must remember that the Russians really started to work on their rocket programs, I would say, 12 or 13 years ago. They took over, of course, a good deal of rocket technology from the Germans.

At that time, the Germans, you may remember, had a plan for an intercontinental ballistic missile. The Germans at the end of the war were seriously considering trying to hit New York. Their idea was very simple: build a big rocket, as big as you can, put a lot of TNT in the warhead and fire them one a day or a couple a day, and after a few weeks there are going to be a lot of holes in Manhattan.

Looking back now, this idea seems to be pretty childish, but, of course, what the Germans and the Russians did not foresee was the development of the hydrogen bomb which makes all this work obsolete.

In my own opinion, if I may put it forward at this point, the Russians are stuck with an ICBM which is much too large to do the job. It will do the job, but it need not be so big. It was designed, as we can see, to carry a very large warhead. We can judge this from the fact that our own ICBM's which have roughly the same purpose and carry roughly the same type of warhead, go over the same distance, of course, are very much smaller.

Mr. MITCHELL. Can you see any connection between that and the fact that the Russians apparently, if their statements are correct, as far as payload, have successfully put heavy satellites in orbit? Do you see any connection?

Dr. SINGER. Simply that they have been working on the rockets for a very long time, and they did perfect them, and they claim these rockets are very reliable now.

Mr. MOELLER. May I ask one more question? We don't want to bore you with personal questions, but do you read these Russian manuscripts in the original?

Dr. SINGER. No, sir; I do not. I do not read Russian. What I do is read the abstracts. If they are interesting, I try to get a translation. The very best ones are in any case translated. Also, the Department of Commerce puts out a very useful weekly publication which cites some of the outstanding Russian scientific papers.

The CHAIRMAN. Mr. Fulton.

Mr. FULTON. You had said that you made your judgment from the record and the statements of the Russians. Of course, that brings up the question of why you have not based your judgment in part on the evidence that U.S. scientific establishments have brought out. It likewise raises the question: Are scientists in universities, such as yourself, who have been interested in the space field, been receiving adequate cooperation from the Federal agencies in making available to you adequate statistics and tapes so that you can make judgments on our programs? You see, we in Congress have said in the law last year that all this was to be made available to scientists such as you, unless it was specifically labeled confidential or secret or classified for the national security.

Now, that brings up the problem of what policy is being carried on by our people in the Federal Government in cooperating with you scientists that are in the private or university field. Would you care to comment on those two points?

Dr. SINGER. I would be very happy to comment on them. They seem to be very important. I cannot make a very general statement about it, but I can say that in this particular case I did not get any cooperation. I tried to get the lunik tapes particularly because I was interested in seeing what scientific results the Russians got out of the lunik shot and that sort of thing.

Mr. FULTON. You say in respect to the Russian lunik or Mechta shot that you did not receive cooperation from the Federal officials in this field, whether military or civilian?

Dr. SINGER. That is correct. I tried several approaches through different channels, for example, through the Air Force Office of Scientific Research; and while the people who I dealt with were very cooperative, they themselves found blockages and they were stopped further up the line.

Mr. FULTON. So that the practical result of it was that you had to make your entire judgment on what the Russians gave you rather than what our U.S. military and civilian scientists in Government had been able to find out, and that would be both.

Dr. SINGER. That is correct.

Mr. FULTON. That is a remarkable situation.

Mr. MITCHELL. Mr. Chairman, if I may on that same point?

The CHAIRMAN. Mr. Mitchell.

Mr. MITCHELL. Doctor, you are, of course, not trying to leave the impression with this committee that there was not some good and justifiable reason for this. There could have been reasons which could not have been revealed to you, which would have been in the interests of the security of this Nation that you didn't receive these tapes. Wouldn't you feel that is true?

Dr. SINGER. I got a little lost on this question. I am sorry.

Mr. MITCHELL. Well, now, Doctor, regarding the questioning of Mr. Fulton to which you responded that you did not receive the cooperation in this particular instance of either the military or civilian space agency. I want to be sure of the impression you intended to convey. You are not trying to say that there was no good and sufficient reason for a refusal for you to get, for example, the lunik tape?

Dr. SINGER. No; I don't think so. I think it is just the tendency to keep things pretty tight. I would suspect that there probably wasn't very much interchange of information between the different agencies who did record the lunik signals. The tendency, I found, is for people who have information just to sit on it.

Mr. MITCHELL. Now, this is pure speculation on your part?

Dr. SINGER. No; it is not.

Mr. MITCHELL. Will you amplify your answer then? That is a very important point. If the agencies are not cooperating as they should, I think that the witness should tell us the information that he has as to why they are not.

Dr. SINGER. This problem merely is much broader, as Mr. Fulton indicated, and it goes even far beyond the field of space research and it is much broader than that. Of course, it applies to the Government as a whole, all of the scientists who are connected in some way with the defense effort. I do consulting, and I have military clearances. In my case, I have had military clearance now since the war, for about 14 years continuously. I also have an AEC clearance now. I have just applied for it. Nevertheless there exists something which is called a need to know. Now, this device is often used as a way of keeping people from finding things out. You see, in order to find out information, you first have to know that a man has it. Well, you can't find that out unless somebody breaks security and tells you. Then after you know he has it, and you go up to him and say, "I want it. I have a need to know." But if nobody knows he has it, you can't get the need to know. In order to get around this thing, somebody has to break security. This has been done all the time. The only reason I find the defense effort works is because everybody is breaking security at all times.

The CHAIRMAN. Breaks in security can be tragically unfortunate, too. Of course, your point is that Government agencies don't release some things that they should. But I can say this, that they release a lot of things that shouldn't be released, also.

Dr. SINGER. If I might extrapolate further, one of my really useful contributions as a consultant to industry is to carry information back and forth. While I cannot tell people what is going on sometimes because I have to establish that the man I am talking to is not only cleared, but he must have a need to know, I can tell him: If you will write to such and such you may be able to get a report which will tell you more about the work which you are doing, but I can't tell you any more than that about it.

The CHAIRMAN. Are there any further questions of this gentleman?

Mr. FULTON. May I make a request that the U.S. Federal agencies at least make as much available to U.S. scientists on their Mechta tracking as the Russians have made available to our scientists on their own tracking. You see, here is a man who is cleared for all security on every basis and says that he cannot get adequate information on a Russian shot. To me that is directly contrary to the act which was passed last year saying that there must be scientific releases to these scientists unless there is a real reason otherwise. In this particular instance the witness says that there is a lack of cooperation, and that there have not been reasons given to him which would put it on any other basis, so I believe that we ought to have a statement from these people as to the policy and whether they are following the statute. I would ask that, Mr. Chairman.

The CHAIRMAN. Did you have something you wanted to say in reference to that?

Dr. SINGER. Well, I only wanted to say that there is nothing personal involved here. I think it is just a tendency for people to sit on data and not release it, even though there is no reason from the standpoint of national security.

The CHAIRMAN. I think we ought to go further into that in executive session. We had some security information given to us yesterday that I think we ought to consider before we go further.

Now, are there any further questions?

Mr. HECHLER. I would like to say, Dr. Singer, that we on this committee really appreciate your expert testimony this morning. In these days when college professors are getting lured away to other things, I certainly hope that you will remain at the University of Maryland and help in this vital function of training our future scientific manpower, because the University of Maryland is very fortunate in having you.

Dr. SINGER. Thank you.

The CHAIRMAN. Dr. Singer, we are very happy to have you here, and I wanted to thank my colleague, Mr. Hechler, for presiding.

Mr. MITCHELL. May I make a comment concerning Dr. Hechler's statement. He spoke of people being lured away. Congress lured him away from a college, too, I might note.

The CHAIRMAN. Thank you very much, Doctor.

We have two witnesses from the Air Force. I would think they want to sit at the table together. One is Mr. R. M. Slavin, the Director of Project Space Track, Air Force Cambridge Research

Center, Bedford, Mass. The other is Dr. Harold O. Curtis, Associate Director, Project Space Track, Air Force Cambridge Research Center, Bedford.

Will you gentlemen have a seat? We are very happy to have you here. You have someone with you.

STATEMENTS OF COL. LINSCOTT A. HALL, USAF, DEPUTY DIRECTOR OF ESTIMATES, OFFICE ASSISTANT CHIEF OF STAFF INTELLIGENCE; R. M. SLAVIN, DIRECTOR, PROJECT SPACE TRACK, AIR FORCE CAMBRIDGE RESEARCH CENTER, BEDFORD, MASS.; DR. HAROLD O. CURTIS, ASSOCIATE DIRECTOR, PROJECT SPACE TRACK, AIR FORCE CAMBRIDGE RESEARCH CENTER, BEDFORD, MASS.; LT. COL. JAMES A. FOX, CHIEF, SPACE SYSTEMS BRANCH, AIR TECHNICAL INTELLIGENCE CENTER, DAYTON, OHIO; AND MAJOR BARNES, AIR TECHNICAL INTELLIGENCE CENTER, DAYTON, OHIO.

The CHAIRMAN. Now, Mr. Slavin, do you want to take the lead?

Mr. SLAVIN. Colonel Hall will.

The CHAIRMAN. All right.

Colonel HALL. I am Colonel Hall, of the Air Force Intelligence, and I have with me Mr. Slavin here, and Dr. Curtis, of Project Space Track, which operates an Air Force Research Center at Cambridge.

I also have with me, sir, Lieutenant Colonel Fox and Major Barnes, of the Air Technical Intelligence Center at Dayton.

Of course, we are here to assist you and members of your committee in any way we can. Unfortunately, until late yesterday, I understood we were to have a closed session, and most of what I have to say is of a classified nature. In fact, I have two short presentations; on Soviet developments in getting man into space, which I believe you, Mr. Chairman, heard last year at another session. I believe you are the only member of the committee that did, however.

The CHAIRMAN. What you are saying will bear repetition. I want the other members to hear it. I will say this, Colonel, we are very anxious on this committee to let the press get as much as they can get, and so I have taken the position, and I think the committee backs me up fully on that, that everything that we can possibly give to the press we do, without, of course, jeopardizing the security of the Nation.

Colonel HALL. Yes, sir; I understand.

The CHAIRMAN. If you will proceed on that basis, we will appreciate it.

Colonel HALL. All right, sir. Mr. Slavin, of Space Track, does have some statements that I think would be pertinent to your problem, and he might also want to address himself to the question raised by Mr. Fulton having to do with the exchange of information between scientists and agencies.

The CHAIRMAN. Would you care for Mr. Slavin to proceed first?

Colonel HALL. Yes, I will ask him to do that.

Mr. SLAVIN. I am Mr. Slavin and I am in charge of Project Space Track. Dr. Harold Curtis, here with me, is in charge of the filter center section of that project.

Space Track has the job for the Air Force of maintaining a catalog of existing and future space vehicles and artificial earth satellites and issuing predictions on their position. In order to do this we have established a filter center at Bedford, Mass., at the Cambridge Research Center, and we rely upon observations made by a rather large variety of observers consisting of the Air Force missile test ranges, some cooperating universities, and observatories, both domestic and foreign, and elements of the other services, Army and Navy, and of NASA, with whom we maintain a free interchange of information.

Specifically, on the 2d of January 1959, following the Russian announcement, we issued a bulletin to our cooperating network, giving the announced frequencies and requesting that tracking and monitoring be accomplished to as great an extent as possible by these stations upon this announced launching.

As a result of this bulletin, we received during the next 30-odd hours a rather large variety of observations on monitoring and tracking of this vehicle.

I have a copy of the observations received here.

Now of this list only two contained tracking data. These two were the Goldstone observation, of which I believe you have heard, and the Stanford Research Institute tracking. The remainder of these were observations of the radiations from the vehicle with no measure of position except as it related to time and no specific tracking.

Upon examination, rather detailed examination, of all of these observations, considering the mass of observations received, their timing and things of this nature, and correlation with the Russian announced trajectory—there was a sequence of announcements made which allowed us to reconstruct a trajectory. From their position announcements, our conclusion is that these observations definitely confirm the Russian announcement that they had a vehicle reach the vicinity of the moon.

Colonel HALL. Are there any particular questions, sir, that you would like further elucidation on?

The CHAIRMAN. You said that he had some observations to make in reference to cooperation with scientists.

Mr. SLAVIN. I wonder if you want to go into that at this time.

The CHAIRMAN. Proceed to that in this general statement.

Mr. SLAVIN. At Project Space Track in particular we maintain a prediction bulletin for all satellites now in existence. These bulletins are published and sent to a distribution list, unclassified, and go to all people who have a request in to us for these bulletins. In general this request is accompanied by some justification for receiving them. There is certainly a transmission cost involved. We cannot just publish them indiscriminately. This justification can be a contribution of further data, which would be a valid justification, a scientific interest in deriving from these bulletins any information would be a justification, anything of that nature, and we do have a free, you might say, a free distribution of the information which we acquire and compute at the control center at Cambridge.

The CHAIRMAN. Do the scientists have access to that information?

Mr. SLAVIN. Yes, sir.

The CHAIRMAN. I mean, whether they are connected with the Government or not. Have you had any complaints reach you as to the lack of cooperation?

Mr. SLAVIN. No, sir, I have not.

The CHAIRMAN. Why would you think that these complaints would come out now at these hearings?

Mr. SLAVIN. In this particular instance, I believe there was a restraint on the part of individuals to release the exact points of data acquired because it was a case of necessity. These evaluations took time. When you make a recording of anything as far away as this shot was, you do not get an absolutely positive, perfectly clear indication of what you have until you have examined your data for days and weeks. For this reason, there was some restraint possibly on some people's part to release these data immediately.

The CHAIRMAN. Now is there any restraint if, for instance, Dr. Singer, from Maryland University, would come to your place and communicate with you, would there be any restraint on giving him information?

Mr. SLAVIN. No, sir. I don't know of any approach that Dr. Singer made to our place in this particular instance.

The CHAIRMAN. I don't know whether he said he did or not.

Colonel HALL. He didn't say so.

Mr. SLAVIN. Not specifically to us, I don't believe.

Mr. FULTON. This restraint then would mean that the evaluation of the data and the tapes could be made by your people, but not by outside people in some instances, because, as you say, it would take a matter of weeks or months for the full evaluation to be made.

Colonel HALL. Yes, sir. I think that the point we would like to make here, Mr. Fulton, is the fact that this was only one type of information which we had to evaluate with many other types of information in order to get something that was really valid in terms of a projection or in terms of an estimate as to how accurate the Soviet statements were. So it did require detailed analysis of many factors.

Mr. FULTON. And the statement was made that your information and data had confirmed the Russsian claim that a vehicle had reached the vicinity of the moon. What does that mean more specifically?

Mr. SLAVIN. I think, sir, that this means that the information which the JPL tracked, I believe, is very nearly a position. The other information which we add to this in this list here gives confirmation to the announced trajectory, the Russian announced trajectory.

Colonel HALL. If I could add to that, I think actually that we feel that there doesn't seem much doubt that this space vehicle got near the moon, but what happened subsequent to that in terms of going toward the sun or in orbit around the sun I think we have to depend entirely upon Soviet announcements and mathematics.

Mr. FULTON. I want to congratulate you, because I feel exactly the same way, and we have had some people already conceding as American citizens that the Russians had been the first people to orbit the sun. There is no information we have available to us in the United States other than what they themselves say from their own data to which we don't have access?

Colonel HALL. That is correct; yes, sir.

Mr. FULTON. On the tracking by Goldstone and the Stanford Research Institute, at what number of miles from the moon did that tracking cease?

Mr. SLAVIN. I would like to pass this to Dr. Curtis, if I may.

Colonel HALL. Is this classified?

Dr. CURTIS. I will have to say I don't know this minute. I would have to look it up, sir.

Mr. FULTON. I am trying to determine how close to the moon it was. We went 78 or 75,000 miles on one of our shots. How far do you think their shot went? Were they 40,000 to 50,000 miles farther than ours? Were they halfway to the moon or three-quarters of the way when our tracking data ceased?

Colonel HALL. I think we feel, sir, they got pretty close to the moon. I believe their estimate was 4,700 miles from the moon. I think we feel that is pretty accurate from what we have.

Mr. FULTON. There is no information then as to the ellipse or any orbit around the moon or anything that occurred after a certain point away, so that we are really just reasoning by inference, aren't we?

Colonel HALL. That is correct, beyond that point.

Mr. SLAVIN. I believe, sir, the velocity data show that there was sufficient velocity to go past the moon.

Mr. FULTON. But nothing in terms of orbit or anything else?

Mr. SLAVIN. Nothing in terms of the definition of the orbit around the sun.

The CHAIRMAN. May I ask you, Colonel, have you been over to Russia to talk with those scientists or in Russia at all?

Colonel HALL. No, sir; I have not.

The CHAIRMAN. Have any of your group been over there?

Colonel HALL. No, sir.

The CHAIRMAN. Could you say this from your own knowledge or from what you have picked up in reading, that the Russians must have very good equipment to handle, for instance, tracking? You say, for instance, they could follow their rocket farther than we could. We rely on them. Apparently they have good equipment, is that right?

Colonel HALL. Yes, sir; we think they have extremely good equipment.

The CHAIRMAN. Do you think our equipment is as good as theirs in tracking?

Colonel HALL. I will let Mr. Slavin handle the specifics on the types of equipment, sir.

Mr. SLAVIN. Specifically the job of tracking requires certainly the knowledge of the frequency on which you are going to track. It requires specifically that you be able to put together equipment far enough in advance to get yourself ready. It also requires equipment of certain band widths of radio frequency reception suitable to the particular mission involved. We have, I believe, as fine equipment as exists for our own lunar shots. When suddenly we are faced with a shot on an unannounced frequency, we are caught rather flatfooted. Do I make myself clear on this point?

The CHAIRMAN. What you mean to say is that the Russians are in on the ground floor, and you come in later on, and your equipment isn't adjusted or adapted to the circumstances, as they have been in advance. Is that right?

Mr. SLAVIN. Yes, sir. I think it is a major achievement that the JPL people got this thing working in the short period of time.

The CHAIRMAN. I think so, too.

May I ask you this: The question comes up on guidance systems. We have been told by the military and others that we think we are ahead on the refinement of these rockets and these missiles, intercontinental ballistic missiles, that we have better guidance systems than the Russians have. Let me ask you this now: You have just testified that the Russians came within 4,700 miles of the moon, and our missile came—what was it, 39,000 miles of the moon? Now would you say from that the Russians' guidance system is not of a superior nature?

Colonel HALL. I would like to correct one thing, Mr. Chairman. First, I said that I tended to believe the Soviet announcement that they came within 4,700 miles of the moon. I don't think we have any evidence that they actually did except a few tracks.

The CHAIRMAN. How close do our tracking stations indicate they came to the moon?

Colonel HALL. I don't think we have any evidence to support this at all.

The CHAIRMAN. Any distance at all?

Colonel HALL. No, sir.

The CHAIRMAN. Will you please amplify?

Colonel HALL. One thing we have done, which is generally well known, is that we have established certain days of the month in the calendar year which is the optimum time for the launching from certain places of the earth. So we know pretty well where the moon should have been at that time. So from the evidence we have, scanty as it is, particularly in the tracking field and emission of signals we pick up, we have the feeling the rocket and the moon were fairly close together at the time the Soviets said it was.

The CHAIRMAN. You don't think they are exaggerating very much on the proximity?

Colonel HALL. Not very much; no, sir.

The CHAIRMAN. From what you have been able to evaluate in the past, would you say the Russians had a very good guidance system in their rockets?

Colonel HALL. I would say their guidance system in their missiles is probably excellent, particularly in their shorter range missiles. I don't think we know what their actual guidance system is in their longer range missiles or how accurate it is.

The CHAIRMAN. When you say shorter range missiles, what range do you mean?

Colonel HALL. 700 or less.

The CHAIRMAN. For 700 or less, they have got a very good guidance system?

Colonel HALL. Yes, sir.

The CHAIRMAN. You don't know about it above that?

Colonel HALL. No, sir.

The CHAIRMAN. Now, therefore, you couldn't say whether or not their ICBM has a superior guidance system or not?

Colonel HALL. It is difficult to say, sir, as to what their actual guidance system is, or how accurate it is. I think we have some information which I would like to go into in closed session that might bear on this point.

The CHAIRMAN. Well, the purpose of these hearings was sort of to evaluate the Russian position scientifically, generally. Do you have any information you could give us on that point in open session?

Colonel HALL. In open session, yes, sir, I could say our information tends to support the belief that they are quite good in their guidance systems, in their propulsion systems, and that for their shorter range missiles that we know a good deal more about them than their longer range missiles, it is quite adequate for military purposes.

The CHAIRMAN. Now they have superior thrust, don't they? They have superior boosters in size than ours, don't they?

Colonel HALL. They have larger engines; yes, sir.

The CHAIRMAN. Larger engines and, therefore, more power to the rockets and the missiles?

Colonel HALL. That is correct.

The CHAIRMAN. On the other hand, beyond 600 or 700 miles in military equipment, you are not so sure about their guidance system?

Colonel HALL. We are fairly sure about it, sir, but we would have to discuss that in closed session in terms of the actual accuracies.

The CHAIRMAN. Of course, it is a very vital question, whether they can hit the moon or go around the moon, but when you come down to the final analysis it is their ICBM that counts.

Colonel HALL. That is correct, but, of course, most of these larger missiles will be carrying atomic or nuclear warheads, too, and the accuracies don't have to be as finite as you have for a smaller payload.

The CHAIRMAN. Do you have available information on the progress of Russian science, generally?

Colonel HALL. Yes, sir.

The CHAIRMAN. The Army Engineers have some maps showing the development of dams and reservoirs and generating power and so forth, all through Russia. I have heard their story, and some of the things that they say are really amazing. For instance, the developments in the Ural Mountain area of Russia. Do you have similar information generally from Russia on the things that especially interest this committee?

Colonel HALL. Yes, sir; I think we do. I think that one of the great problems we have in the American community is recognizing the technical advances that have been made in the Soviet Union in the last few years, particularly since the war.

The CHAIRMAN. Generally, what is the nature of the technical advances there, as you see them?

Colonel HALL. Well, if we could take the field you are particularly interested in, in terms of missiles and space, I think one of the things that they are considerably ahead of us in the missile field is in experience. When you realize they have been actively involved in this missile business not only during the war with their rockets—I mean actual artillery type rockets, unguided—but also beginning about 1947 or 1948 they have been very active in the guided missile field, and this puts them considerably ahead in terms of experience.

I think that in terms of the specifics of the guided missile field the United States has made tremendous progress, but they still have a way to go probably to catch up with the Soviets in terms of guidance and propulsion, which are the two major factors in the missile field.

The CHAIRMAN. Do you think, then, we are behind the Soviets in guidance?

Colonel HALL. I would say that is probably true; yes, sir.

The CHAIRMAN. Now, in the gadgetry, generally, that is used in the missiles and rockets, are we ahead of the Soviets from your information?

Colonel HALL. Well, if you want to get down to terms of miniaturization or something like that, I think we are probably considerably ahead of the Soviets in miniaturization, but do they need it when they put up sputniks in terms of 3,000 pound weights when we put up something the size of a grapefruit? You have a different mission and concept of operations here. I think you have to realize the Soviets have one program and our is not necessarily adopted in the same fashion. So in certain aspects of the missile field we are ahead, and miniaturization is one of them.

The CHAIRMAN. Whereas they are ahead in power?

Colonel HALL. I don't think there is any question about it, sir, that their engines are larger and that they have probably made considerable developments in ceramics and propulsion that we probably haven't caught up with as yet.

The CHAIRMAN. You are from this research center at Bedford?

Colonel HALL. Not I, sir. I am here in Washington.

The CHAIRMAN. You are from the Air Force?

Colonel HALL. Yes, sir.

The CHAIRMAN. Have our people been able to get into the range that they use over there or any of their factories that make these missiles, make these rockets?

Colonel HALL. Yes, sir; we have information about their test ranges, which I would much prefer to talk about in closed session.

The CHAIRMAN. Well, I will not pursue that any further.

Are there any questions?

Colonel HALL. Mr. Slavin says he wants to add something.

Mr. SLAVIN. One thing I want to add. This happens to be a field that I have been associated with, even as Dr. Singer was, in the capture of the V–2's with the Air Force's upper atmospheric research program. I am interested in the experiments that have been performed in these problems. I have seen some of the Russian equipment and feel again that while ours is pretty, they are doing the same job with a little bit more brute force, and in the miniaturization field if you have the brute force you don't need the miniaturization.

The CHAIRMAN. So if we are behind, we are not hopelessly behind at any rate?

Colonel HALL. No, sir.

Mr. SLAVIN. They have been doing the same thing.

Mr. RIEHLMAN. Can you give us in open session more information on which you base your statement that Russia has better missile guidance than the United States?

Colonel HALL. Not very well, sir.

Mr. RIEHLMAN. It would have to be in closed session?

Colonel HALL. I think it would be more meaningful if it were in closed session, sir.

The CHAIRMAN. Are there any further questions?

Mr. FULTON. Could I pursue that a little further? When you are speaking of guidance, do you mean inertial guidance? Do you mean command radio guidance and ground control?

Colonel HALL. I would say primarily in the inertial guidance systems.

Mr. FULTON. Obviously we are the ones who are tremendously further ahead in the ground station control and in the network of telemetric and tracking stations all over the world; aren't we?

Colonel HALL. I would say that is probably true.

Mr. FULTON. As a matter of fact, on these shots of Russia the type of guidance that you are speaking of, that they are ahead of us possibly, is actually only the guidance at the time of the gravity turn, when the missile becomes a ballistic; is that not right?

Colonel HALL. That is true; yes, sir.

Mr. FULTON. Because we have no real evidence that they have any radio transit control of the Mechta or the lunik after it took the gravity turn?

Colonel HALL. That is correct.

Mr. FULTON. As a matter of fact, the pattern of the tracking showed a ballistic course; did it not?

Colonel HALL. Yes, sir.

Mr. FULTON. On both Goldstone and on the Stanford Research.

Colonel HALL. I would like to point out, though, here, Mr. Fulton, that the tracking information is extremely scanty, but what we do have confirms that.

Mr. FULTON. That is what I am coming to. Then at the time that the tracking ceased the missile was in the earth's gravitational field and the velocity was less than escape velocity; was it not—less than 7 miles a second?

Colonel HALL. I don't know that. Let me ask one of my people who can answer that.

Colonel FOX. No, sir. I think the best we can say is that the velocity when observed appeared to be adequate, but the position that Goldstone has, I think Mr. Slavin is in the position to answer rather about trajectory.

The CHAIRMAN. Would you give your name to the reporter?

Colonel FOX. Lieutenant Colonel Fox, of Air Technical Intelligence.

Mr. SLAVIN. At the time of the tracking at Goldstone and at Stanford, the tracking, I would say that the lunik was in the moon's gravitational field, Mr. Fulton.

Mr. FULTON. That is right, but it did not have a sufficient velocity to escape the earth's gravitational field at that particular time when the tracking ceased.

Mr. SLAVIN. I have no evidence on that.

Colonel HALL. I don't believe we have any evidence to support that statement, Mr. Fulton.

Mr. FULTON. So, therefore, the statement by the Russians that they have a new planet orbiting the sun, as far as we are concerned is wholly without scientific data or confirming information; is that not right?

Colonel HALL. It is based on the Soviet announcements, that is correct, and mathematics. That is all.

Mr. FULTON. As a matter of fact, we know nothing on the lunik or the mechta, whichever you want to call it, announced on January 2, 1959, other than it was heading in the direction of the location of the moon and would have gone in that direction had it continued on from the time that the tracking ceased?

Colonel HALL. I wouldn't say we know nothing. I would say that the conclusion that you came to is correct, but we know a great deal more about the information which we would like to discuss with you in closed session.

Mr. FULTON. Have any of our computers come up with the terminal point on the circumference of the orbit of the very extended trajectory? Do you have any terminal-point computations?

Colonel HALL. Do you mean estimate around the sun?

Mr. FULTON. On what the terminal point would be on the ellipse. If it is a very extended narrow ellipse, where would the terminal point have been when you had completed your tracking?

Mr. SLAVIN. From the meager tracking data, I think it is fair to say that a slight error in this tracking data could produce a tremendous change in the ellipse that the satellite would have actually entered into. Out of all of these possible ellipses possibly one out of a hundred might have impacted the earth again. The others would have become orbits of the sun.

Colonel HALL. Based on mathematical projections.

Mr. SLAVIN. This is based on the trackings.

Mr. FULTON. If they had been in orbit in the sun and the velocity when you finished tracking was greater than the escape velocity from the earth's gravitational field, and taking into account the law of its slowing down as it went.

Mr. SLAVIN. It is a very complex situation, sir. As far as we know from the observations, the thing was within the moon's gravitational field at the time of the tracking observations, to which we are referring, and apparently in a position and with sufficient velocity to escape that field and go past the moon, at which point it then becomes an object subject to a large variety of ellipses, depending upon the precise value of direction of motion and velocity.

Mr. FULTON. And actually it would be then a determination of one type of an ellipse with the influence of the moon's gravitational field and would then pick up another kind of an ellipse because of the sun's gravitational field; is that not right?

Mr. SLAVIN. In other words, in the travel of this rocket you could almost look at it as being three ellipses, first in the earth's, and then the moon's and then having escaped from that into the sun's.

Mr. FULTON. Because of our lack of information and the ceasing of the tracking data at a point that is removed in lunar space from the moon, we are unable to say that this particular lunik or mechta entered any one of those ellipses permanently?

Colonel HALL. That is correct.

Mr. FULTON. We are just saying that there is a whole astronomical number that it might have come into, because our tracking equipment or our tracking data is not accurate enough to predict what did happen; is that not right?

Mr. SLAVIN. I think we can come pretty close to saying that this did not come back into the earth's gravitational field.

Mr. FULTON. I will agree with you on that, but you see I am projecting it further and saying that we cannot concede with the Russians that they have put an extra planet in orbit around the sun, as we do not have the scientific data nor the tracking data to confirm such a judgment.

Colonel HALL. I think one possibility has not been mentioned here, and that is, it may have impacted into the sun.

Mr. FULTON. I tried to say that yesterday and was laughed to scorn, so I am glad to have you join me on that.

Colonel HALL. I think it is a good likelihood.

Mr. FULTON. Yesterday there were three scientists here who thought it was impossible.

Colonel HALL. I am not a scientist. I am just an intelligence officer. That is what my feeling is.

The CHAIRMAN. Are there any further questions? If not, do we have any witnesses now that we can hear in open session?

Mr. BERESFORD. No, sir.

The CHAIRMAN. We have to go into executive session at this time.

(Whereupon, the committee proceeded in executive session.)

SOVIET SPACE TECHNOLOGY

MAY 14, 1959

House of Representatives,
Committee on Science and Astronautics,
Washington, D.C.

The committee met at 10 a.m. in room B-214, New House Office Building, Washington, D.C., Hon. Erwin Mitchell presiding.

Mr. Mitchell. The committee will be in order.

This morning we have Dr. Hans Ziegler, of the Army, who, as I understand, will be the first witness. Do you have a prepared statement?

Dr. Ziegler. No; I do not.

Mr. Mitchell. Please be seated and proceed in any manner that you choose.

STATEMENT OF DR. HANS ZIEGLER, DEPARTMENT OF THE ARMY; ACCOMPANIED BY WILLIAM H. GODEL, ADVANCED RESEARCH PROJECTS AGENCY

Dr. Ziegler. I am not quite sure how to proceed, Mr. Chairman. Am I here for particular questioning, or are you——

Mr. Mitchell. Of course, you know the purpose of your appearance here this morning. Do you have a preliminary statement you would like to make before the committee members direct questions to you?

Dr. Ziegler. Yes.

I understand there are two questions placed before me. No. 1 was what kind of equipment and facilities we have in Fort Monmouth which can be applied to space probe tracking and, No. 2, what kind of activities did we conduct during the period of January 2 through January 5, during which the lunar probe of the Russians was supposed to be traveling in space?

Mr. Mitchell. I think the committee would appreciate you directing your first statements to those two questions.

But I think, first, if you would—I anticipate Mr. Miller would probably ask you this anyway—give us your background, your training, and the work that you have done heretofore in this field, Doctor.

Dr. Ziegler. I received my training all in Germany at the University of Munich. I have a bachelor's, master's, and doctor's degree in engineering of the Technical University of Munich.

In electrical engineering I have served for 2 years as assistant professor at the University of Munich in electrical engineering and then went into the electronics industry. This was during the period

48438—59——8

of 1937 through 1945 in which Germany was preparing or engaged in World War II. My major activities were military electronics, particularly proximity fuses, guidance electronics for missiles, and associated equipment.

I was invited to the United States in 1947 and joined the Signal Corps, with which I am still employed, originally as scientific consultant, later as Assistant Director of Research. Presently I am the director of a newly established astroelectronics division which handles the entire work of electronics in space, as far as the Signal Corps is concerned.

I may say we take considerable pride that out of that activity three successful major satellite payloads out of a total of seven successful U.S. launchings have come.

We have developed the small 4.6-inch vanguard satellite. We have operated and developed the score satellite equipment, equipment launched in cooperation with the Air Force in December 1956 and also provided the vanguard cloud cover equipment which was launched in February 1959.

Mr. MITCHELL. All right, proceed.

Dr. ZIEGLER. As far as the facilities in Fort Monmouth are concerned, I would like to point out, first, that these facilities are not part of an operating facility as such for tracking purposes. They have been established and initiated for other purposes. Those purposes are research and development in wave propagation, ionospheric parameters, and overall worldwide communication parameters.

The basic bulk of this equipment and facilities exists in the form of a very large variety of different types of antennas—antennas with different frequency ranges, different directional characteristics, and different polarization characteristics. Many of them are of the so-called low-gain type, similar to television applications in commercial use, and some of them are of the very high-gain type, like parabolic dishes.

Among those we have a 60-foot parabolic dish and a 50-foot parabolic dish. I trust the committee is aware that one of those installations had been used in 1946 to make the first contact with the moon, which actually we can consider as the first step of man's endeavor into space.

In addition to those many types of antennas, we have associated high sensitivity and low-noise receiving equipment and recording equipment, and we have located this equipment in particular areas which are not influenced by electronics interference. One big area is a 400-acre tract in Deal.

As has probably been mentioned quite often, Deal is one of the tracking stations. We have the big dishes at the Belmar and Collingwood sites at Belmar.

And we have, in addition, special finding equipment at an extra test site some 10 miles from Asbury Park in Collingwood. Those are the major equipment, antenna, receiving equipment, recording equipment, and direction-finding equipment.

In doing our research work for propagation purposes, which means providing parameters to improve overall military and civilian communications, we have found, as soon as satellites appeared, that we can combine our scientific endeavors with practical tracking results.

We have made available to the tracking stations, principally for the vanguard and space tracks and to all the special computing centers established for special projects and space probes the results which came out of our research work and applied to tracking.

Since we were so very successful, 6 hours after Sputnik I was announced, we have continued to operate those stations on a 24-hour around-the-clock basis.

We have never interrupted it and we are operating on a number of frequencies. So if, for instance, the Russians would launch something again on 20 to 40 megacycles, we would have it whenever they do it.

We have done one other thing: We have recorded all of those observations, so that there is a complete library in existence which covers all satellite orbits, United States or Russian, whichever have flown over the United States.

They are all recorded and available and numbered and can be reproduced whenever necessary.

We have, as I said, a number of possibilities to do something in research, which at the same time applies and gives results for tracking but there really are certain limitations.

If we talk about observation of space probes, we mean the high gain antenna like the 50- and 60-foot dishes. Those dishes have in the present state of the art one big disadvantage.

No. 1, they are not very applicable for frequencies below 50 megacycles, because the dish size has to have a certain relation to the wave lengths and the wave lengths are too large for those small dishes.

So we are only effective on higher frequencies.

No. 2, the dishes are only designed for a certain frequency and it is difficult to change the frequency to another frequency. They need different feeds, different adaptors. It takes time to do this, and even if you have some available, money has not been available to make a whole set from say 50 to 1,000 or 1,500 megacycles, as would be decided.

But things are being repaired. If ever an unannounced space probe comes up and we do not know the frequency well in advance, it is very difficult or was, in the past, difficult to adjust the sudden high to a new requirement, and when the Russian lunar probe was announced at 180.3 megacycles, there was just no such feed available and this was one of the difficulties that JPL people encountered. However, they were able to set up some makeshift device.

Now, in coming to our activities during the particular days in question, I would like to say the following: We had been alerted by Space Track—I indicated already that we are in continuous contact with all those tracking organizations and provide them with data and they, in turn invite us to particpate in the various events—we were alerted that something of that nature may come up on the 1st or 2d of January but the event hit us at a very bad time.

You remember, I indicated already we had the Score project, the first communication satellite launched on the 1st of December, and we were in the process of having this important experiment up. We had our crews in five ground stations in California, Arizona, Texas, Georgia, and Florida, and we had also our own tracking station or observation stations in Deal completely occupied with this particular very important event and all working on 108, 132, and 150 megacycles. So, we were not ready to go on other frequencies.

We had nothing for the dish, to do something in a hurry, and possibly if we had those crews which were out on the other sites, we could have produced makeshift devices as JPL did. But we did not even have the people who were out on the other sites.

Mr. MITCHELL. If I may interrupt you—did you have on the sites the necessary equipment to convert to 180.3 megacycles?

Dr. ZIEGLER. Pardon me?

Mr. MITCHELL. Did you have the necessary equipment with which you could have, in a reasonably short time, converted as they did at Pasadena?

Mr. ZIEGLER. No. We did not have any prepared feeds available, but probably we could have, with all the people available, gone to the workshop and made something in a hurry which may have worked in a way similar to the makeshift device at JPL.

But we also had a very bad handicap due to the fact we were engaged in the other project which we felt was very important, of course.

Nevertheless, we were very capable of going on the other announced frequencies, which were those in the area of 20 megacycles, 19.993 through 19.997.

After our people received the alert, we went on those frequencies, and the Deal station has recorded 40,000 feet of tape recordings on this particular event through the entire time from January 3, 0049, eastern standard time, through January 5, 1735 eastern standard time.

We have a complete set of recordings on those frequencies, and as a matter of fact, we have received, as was announced, signals on those frequencies.

It is somewhat difficult to identify those signals clearly as coming from a certain source.

If they are coming from ground stations which are well established and have legitimate origin, they usually are identifying themselves by call letters, which sometime during a 2-day run, would appear.

There has never been any call letter on those frequencies.

On the other hand, if they are coming from something which moves in space or into space, we expect Doppler effect readings, changes of frequencies. There were changes of frequencies, erratic during the 2 days, but we were not able in any instance to track long enough on those signals to get a definite type of a Doppler effect.

Besides, one has to be very careful of signals on this particular frequency. I would like to make a very crude comparison.

On a frequency of this kind, we have an effect like porpoises in the ocean. The frequency jumps in and out of the ionosphere and you do not know where they come from. You may not get the Doppler effect with regard to your own position on the earth and the moving object.

You may get it only from the moving object to the entire envelope of the ionosphere of the earth. So it is very much more difficult to distinguish the Doppler effect on this type of frequencies.

We have had frequency shifts but we were not able to clearly identify the Doppler. The signals that were coming in and out were not very steady, and I would say, if they were the only observation on this particular space problem, that they would certainly not be conclusive, but in conjunction with the other observations they may have a significance.

They are certainly in the area of frequencies that have been quoted in which the Russians would transmit, and they have appeared in the time frame which would be expected. They have had no characteristic which would contradict that they could have originated from a space program.

I think this is all we can conclude from those signals, and as I said, there are still available 12 reels of half-inch tape with 40,000 feet overall contents. They still can be evaluated if any more details would be necessary, or possibly have been overlooked.

I think this is all I am able to state at this time.

Mr. MITCHELL. Are there any questions at this point of Dr. Ziegler by any member of the committee?

Mr. FULTON. The question comes up on the parameters you have established that you have spoken of. Would you please be more specific on those?

Mr. ZIEGLER. As a result of our wave propagation research?

Mr. FULTON. Yes, at your particular installations. What parameters have you specifically established?

Dr. ZIEGLER. For this particular night or——

Mr. FULTON. No, generally.

Dr. ZIEGLER. Well, we are establishing the overall basis for operating long-distance communication circuits, radio circuits.

Mr. FULTON. Give us the limits; that is what I am asking, of your parameters that you are setting up for these tests.

Within what limits have you set up parameters, generally? I am going to ask you the same thing on the specific type of space program.

Dr. ZIEGLER. I am afraid this is somewhat difficult to answer exactly.

If you think of how far can our equipment reach—is that what you are after?

Mr. FULTON. Various elements. I am trying to find out what you mean by establishing parameters.

It was indefinite to me. So I say, please make it more definite so we can understand what it is.

Dr. ZIEGLER. This is a mutual understanding between the transmitting power of the space probe and the ground equipment sensitivity and we have tracked as far as U.S. probes are concerned. We have been able successfully to track some through their entire orbits; so if the Russians or anybody uses similar comparable power outputs in the space device, we would be able to follow it. Is that coming close to your question?

Mr. FULTON. Yes. You see, you are the person who testified you were setting up certain parameters. I am simply saying, will you please be more specific as to your general parameters, and also how those parameters would apply in this particular instance?

Dr. ZIEGLER. I am afraid this is very hard to express in clear terms because we have the ground equipment as sensitive as the state of the art presently permits it.

We believe there is hardly anything more sensitive available. As far as dishes are concerned, we are going into the 60 foot. Other people have larger dishes. If they have an 85-foot dish, they can add on more sensitivity due to the meter gain.

Mr. FULTON. What wavelength do you take care of with that type of a dish?

Dr. ZIEGLER. This can be used for any kind of wavelength above 50 megacycles, as long as you have the proper type of feed, but for the longer wavelength and for the higher frequencies it would be more effective than for the lower frequency.

Mr. FULTON. Your particular type of installation is not very effective then, for the particular wavelength that the Russians have chosen?

Dr. ZIEGLER. Not for the 20 megacycles, but for the 182, if the Russians used it, if we had feeds available for this particular antenna, which we have planned and which we will receive.

Mr. FULTON. When will you be able to get those feeds?

Dr. ZIEGLER. I think we will have complete coverage from 50 to 1,000 megacycles by midsummer of this year.

Mr. FULTON. My point was, I wanted the parameters you set up and I wanted to engage those with what the particular lunik shot brought up, to see whether you were adequately staffed on equipment and personnel.

Dr. ZIEGLER. As far as the basic facilities are concerned, we are, except there are additional facilities which we already have in work to make them adaptable to the particular frequencies which have been used. For instance, we have already made the 160-megacycle feeds which we use on the U.S. lunar probes and which have been used already.

Mr. FULTON. At the 182–183 level that the Russians have at least used once, what kind of a receptivity will you have?

Dr. ZIEGLER. There will be very, very effective receptivity, and it will be hardly significantly different from an 85-foot dish which has been used by the Goldstone tracking site, which was obviously able to get the signals very well.

Mr. FULTON. Then, in evaluating your judgment, you have said at this particular frequency level, that these waves do a skip-jump in the ionosphere and some come through and others bounce. Others go right around the whole earth's magnetic field so you get a receptivity as if it were on the whole field?

Dr. ZIEGLER. Yes.

Mr. FULTON. In your judgment, is that accurate enough for you to make any decision, based on your own equipment, as to whether the lunik, or the Mechta did what the Russians claim it did?

Dr. ZIEGLER. Based upon the results alone, I would say we have no right to make any final conclusions, but they may support, as they stand, the other information. We could not say the signals came from a certain direction, because this would not be applicable to this particular frequency.

Mr. FULTON. So that actually you just know that there were certain signals that came from outer space in the lunar area, but you cannot tell the direction, what it was nor can you tell by the Doppler effect what the receding velocity was?

Dr. ZIEGLER. That is correct. We can only say we have received signals in the frequency range which was announced, that what we heard was in the time frame and that we have not been able to identify them as coming from any specific ground source. This is about all we can say.

Mr. FULTON. And, as far as you know, the signals could have been coming from either a parabola course, a closed ellipse, or a circular orbiting course, and you would not have known the difference.

Dr. ZIEGLER. Yes.

Well, on the second orbit we probably would have none, because it would have been repetitive and it would have shown a certain pattern but any other details of a trajectory could not be detected.

Mr. FULTON. Even on an elliptical course you could correlate the repetition.

Dr. ZIEGLER. Probably.

Mr. FULTON. But as soon as it got beyond the limit it went beyond the capacity of your instrumentation and receptivity; right?

Dr. ZIEGLER. Right.

Mr. FULTON. You were not able to distinguish any Doppler effect at all?

Dr. ZIEGLER. There was Doppler on it, but it was never long enough to make any reading upon which you could actually take positive values.

There was some changes on frequencies, but it is very difficult to decide, if you do not get a long observation. If there was an instability of this particular oscillator, it might have shifted one way or another, or is it due to a motion in the trajectory. So we did not come to a conclusion on the nature of the signal.

Mr. FULTON. So, from your particular receptivity and the data you have assembled as well as the study of it, you have stated your own findings would not be conclusive on the Mechta or the lunik problem which was announced by the Russians on January 2, 1959?

Dr. ZIEGLER. Alone, as they stand, it would not be conclusive.

Mr. FULTON. If they are not, then, in and of themselves conclusive, they can only be added to other evidence, being cumulative evidence, as distinguished from evidence which establishes the course or the direction or the distance?

Dr. ZIEGLER. That is correct.

Mr. FULTON. Then you have ended up by saying that it may have an effect, when you combine your data with other data, because you are simply filling in certain interstices and looking for a correlation and, secondly, you were really trying to help establish that it was not something which negated what the Russians said. It that not about what you said?

Mr. ZIEGLER. Yes, sir.

Mr. FULTON. Is that right?

Dr. ZIEGLER. Yes.

Mr. FULTON. From your own data and your studies, would you please tell us if you can, at any point of time to another later point of time, the course, the acceleration and velocity, as well as any possible terminal point of the lunik rocket?

Dr. ZIEGLER. I am sorry, Mr. Congressman, it would be impossible to get this from these results.

Mr. FULTON. So that really what you are saying, in ordinary language, is that you are not able at any point in time, or between any two points in time, to say where it was going, how fast it was going or where it might land?

Dr. ZIEGLER. No.

Mr. FULTON. All right.

Mr. MITCHELL. Doctor, you say from other information made available to you that the information you received from the signals is consistent with the fact the lunik probe was launched at the time and followed the course the Russians announced; is that not correct?

Dr. ZIEGLER. Yes.

I would like to make this clear: I said the frequencies which were announced and which the Russians would transmit in that time frame in which it was expected, coincided with out observations.

Mr. MITCHELL. The signals you received, you made it clear, were such that you could say nothing positively. Now, can you account for these signals as coming from any other source?

Dr. ZIEGLER. As I indicated already, they had no identifications themselves, which they should have if they had come from any amateur or commercial source. They did not have any such identification.

And I am sure, and probably this is part of something which should be discussed in the executive session, if there had been a source of that kind some place else, it would have been detected.

Mr. DADDARIO. You said you took 40,000 feet of tape recording at a certain time from the 3d of January until the 5th of January.

Dr. ZIEGLER. Yes.

Mr. DADDARIO. And the information on these tapes, some of which still has to be evaluated, coincides with other information which you have been able to gather together since.

As you eliminate the various possibilities and as you are able to take into consideration the frequencies and everything else involved, do you not come to certain conclusions which are definite and positive conclusions, rather than negative?

Dr. ZIEGLER. I do not think they are positive or negative.

Because what you receive here is 5 minutes of information during a certain interval. Then it blacks out.

Then you get a few minutes later another signal. And this is a typical arangement which you get, due to the fluctuating of those frequencies in the ionosphere, and it is never enough to give you a whole half hour of information which could be evaluated under Doppler.

Mr. DADDARIO. I understand that, but you had specific information to go on, certain information from the Russians as to megacycles; did you not?

Dr. ZIEGLER. Yes; this is the only information we had, in that frequency range.

Mr. DADDARIO. As I understand it, 6 hours after Sputnik I went up until today, you have had a 24-hour alert in this area?

Dr. ZIEGLER. Yes.

Mr. DADDARIO. On this occasion, the lunik rocket went out and finally went out of range somewhere.

Dr. ZIEGLER. Yes.

Mr. DADDARIO. Except for that particular period, have you ever received signals of the same type on the same frequency?

Dr. ZIEGLER. No.

Mr. DADDARIO. Therefore when you begin eliminating, and you begin putting all of this information together, do you not have to draw a conclusion that it cannot have occurred otherwise—that this information could not have been transmitted except from this particular rocket and could not have come from nowhere?

Dr. ZIEGLER. That is the only conclusion you can make.

As I said, it has never been observed in the same frequency range, but since it was a frequency which was announced and fitted into the time frame, we said it could be that signal.

Mr. DADDARIO. It was the only time in which this particular information was recorded?

Dr. ZIEGLER. Right.

Mr. DADDARIO. Therefore, is that not a positive conclusion, a positive result, eliminating everything else, that the signals could only have come from that source?

Dr. ZIEGLER. Well, if we would exclude that somebody plants a transmitter somewhere and makes this signal, this would be another solution. But, as I said, this has obviously not been observed, so it is not an exclusive conclusion. You could not say that since this signal was in fact never heard before and it had been heard at that time, you cannot say it can be only from space. It can be from some place else.

Mr. DADDARIO. If we could explore that a moment, do you mean if somebody transmitted from some place on earth——

Dr. ZIEGLER. That could happen.

Mr. DADDARIO. This particular wavelength?

Dr. ZIEGLER. Or several places on earth.

Mr. DADDARIO. Or several places.

Dr. ZIEGLER. They probably would have to use several places. One would not be enough.

Mr. DADDARIO. Would it then create the same kind of result?

Dr. ZIEGLER. It could.

Mr. DADDARIO. And there would be no way to check it so far as the Doppler effect is concerned, and the reaction occurring on your recording?

Dr. ZIEGLER. No; you can insert an artificial Doppler and they could change the frequency. If somebody wanted to be real facetious, they could do a lot of things.

Mr. DADDARIO. This is important because we are trying to draw specific conclusions, and, as I understand your statement, it could have come from a rocket which went beyond the moon.

Dr. ZIEGLER. Yes.

Mr. DADDARIO. And it also could have been simulated in other ways from a broadcast of this type from one or several places on earth during this particular period of time?

Dr. ZIEGLER. Yes.

Mr. FULTON. Would the gentleman yield?

Mr. DADDARIO. Yes.

Mr. FULTON. Did any orbiting satellites with signals of the same frequency just suddenly go off, because you could simulate the signals and the Doppler effect? The point I am making is: We did not know what this particular thing was doing from our own U.S. data and tapes.

Secondly, I would disagree with you that it went beyond a range. All we know is that the signals stopped at a certain point. We do not know whether they went beyond the range and that is why I will not give the Russians the credit.

I will not give them credit for going beyond the moon unless I have some information.

Mr. DADDARIO. We are not in disagreement. I am just trying to find out what these people found.

Dr. ZIEGLER. Going back to the theory it can be created by some artificial arrangement, I am afraid this would have been discovered by other arrangements than ours if there were some transmitters on earth which transmit this signal of a particular nature.

I do not think we can go into these details at this particular session.

Mr. DADDARIO. I understand there can be checks and doublechecks insofar as particular agencies are concerned. But if it was the intention of the Russians to simulate a location—and they were very secretive about it and spread their transmitters around in certain areas—they could then, during this particular period of time, have simulated this particular condition and we would not have been able to determine whether it came from these sources or from outer space.

Is that what you are saying?

Dr. ZIEGLER. So far as our observations are concerned, that is correct. But there might have been other observations which could have identified that.

Mr. DADDARIO. You have no way of knowing whether any of these things could have been determined by somebody else or whether there were other ways of checking.

But from your own observation, as I understand it, and from the information you have on hand, this signal could have come from a satellite which did what the Russians say it did, or it could have been simulated.

Dr. ZIEGLER. Yes.

Mr. FULTON. May I follow that a minute?

Mr. DADDARIO. Yes.

Mr. FULTON. You have seen no tapes or information of a U.S. source, other than Russian material, which solved that dilemma for you, on the basis of which you could make a clear judgment?

Dr. ZIEGLER. I am afraid this is something which we cannot discuss in this session.

Mr. FULTON. All right.

Mr. GODEL. Perhaps if I answered "Yes." The tapes have been seen by Dr. Ziegler.

Mr. MITCHELL. Were the signals you recorded sufficiently clear? There has been some testimony they had a peculiar sound.

Did you get a very distinct sound from your recording of these signals?

Dr. ZIEGLER. There are indications that there were several types of modifications that were keying on it and there was particular sound on it.

This has been recorded at our station as well.

Mr. MITCHELL. Did you hear the recording that the Russians made that supposedly came from the lunik launching?

Dr. ZIEGLER. I do not think that I did.

Well, I have heard this but I do not think I could detect any real—this is very difficult to compare, because we all use different insertion frequencies in order to beat the tone. So, in one case you get a completely different acoustical phenomenon to what you get in the other. So this would not be easily compared.

Mr. MITCHELL. Are there any other questions?

Mr. KARTH. Do you, personally, know any scientists who are now working for the Russian Government?

Dr. ZIEGLER. No, I do not.

Mr. KARTH. Or any electrical engineers?

Dr. ZIEGLER. As a matter of fact, I do not know any individual at all who is presently working for the Russians.

Mr. KARTH. Are you at all familiar with any of their technology in this field of space?

Dr. ZIEGLER. I was in Moscow for a period of 2 weeks last year on a visit and I got as much as they permit you to see there.

Mr. KARTH. Are you familiar with the article that appeared in True magazine recently?

Dr. ZIEGLER. Yes.

Mr. KARTH. Do you agree with the conclusions drawn by Mr. Mallan that they were 10 to 15 years behind us in technology in this field?

Dr. ZIEGLER. I am afraid I have no absolutely sure basis to contradict this, but my own observation would tend to somehow contradict.

Mr. KARTH. In a few words, if you can, Doctor, tell us what your opinion is of Russian technology in this field. How does it compare with ours?—if you can tell us in open session.

Is it fairly adequate in comparison with ours; is it not adequate; and in what way is it not adequate—if it is?

Dr. ZIEGLER. I have the impression that it is comparable with ours. There may be certain areas—and I believe the article in True magazine is correct—for instance with regard to component size, they have not placed as much emphasis on miniaturization. They are using rocket designs of conventional type in many cases. But, on the other hand, they have had at their disposition payloads which did not make it so critical, at least at the beginning, as it did for us, to resort to the smallest and lightest components, in order to get equipment together.

So they may be somewhat less advanced in this particular area. I have seen some of their satellite equipment. It was said to be the one they actually fly—but you never can tell—which was built the same way as we would probably build for vehicular ground equipment—real heavy and with ground components.

Mr. KARTH. Are they capable of a lunik?

Dr. ZIEGLER. I would expect they would be.

Mr. KARTH. Do you think they are capable of making a moonshot? Mr. Mallan made quite a point of this.

He said they were not and would not be for many many years.

Dr. ZIEGLER. I think they would be, but this is my personal opinion; I do not have an absolute basis.

Mr. KARTH. No; from what you know and what you have seen.

Dr. ZIEGLER. From the overall impression, which is a limited one, I would expect they would be able to do this.

Mr. KARTH. Being an expert in the electronics field, Doctor—when you looked at their electronic equipment, did you feel it was fairly good equipment, about on a par with ours, or did you feel it was obsolete as suggested by Mr. Mallan?

Dr. ZIEGLER. I think this is very difficult to say. They probably did not let us see the latest developments anyway and this was obviously one of their tactics to visitors, which has happened in many other fields. They show a certain state of the art, which is not the latest one, and they also usually make it a point to let you talk to people who are not exactly on the frontline of development. They may be very reputed scientists but may not be on the frontline of development.

Mr. KARTH. Did you talk to any of the scientists Mr. Mallan talked to?

Dr. ZIEGLER. Yes. But I was surprised he did not get in contact with one of the most important ones and that is Professor, or Major General Blagonravov, who is actually leading the space program.

Mr. KARTH. From those you talked with, who were the same people Mr. Mallan talked with, did you draw the same conclusions from your conversations as he did?

Dr. ZIEGLER. There is only one person whom I talked to, who is mentioned in this book, and this is Masevich, who is concerned with the tracking, and I have not got into the optical aspects of tracking. This is not my particular interest, and I do not know if the equipment which was shown to him is actually the latest or the only ones the Russians have.

It is very difficult.

Mr. WOLF. I just wonder if Mr. Mallan got the "snow job" or the American people are getting the "snow job," or both.

Dr. ZIEGLER. Probably there is somehow a compilation of those possibilities.

Mr. KARTH. Did you see any receiving equipment that the Russians had or are using?

Dr. ZIEGLER. I have seen the ionosphere station in which they use their own developed equipment.

Mr. KARTH. What did you say?

Dr. ZIEGLER. I saw the ionospheric sounding equipment.

Mr. KARTH. And your reference to it was what?

Dr. ZIEGLER. They used equipment they produced in the country themselves and they also used equipment which they import from Czechoslovakia, and from Eastern Germany, and from Hungary. This equipment was very well built and very well comparable with our own equipment.

Mr. KARTH. Did you answer my question as to whether or not they were capable of making a moonshot?

Dr. ZIEGLER. Yes, I did, and I said my personal feeling was they could.

Mr. DADDARIO. Will the gentleman yield?

Mr. KARTH. Yes.

Mr. DADDARIO. You say your personal feeling is they could. What is that based on?

Dr. ZIEGLER. On Sputnik I, II, and III.

If we can trust at all what is known about this, but what we cannot check ourselves as correct, there is no reason we should not be able to do that.

Mr. DADDARIO. Well, if you take Sputnik I, II, and III, is there a possibility that there is a range beyond what they can do in Sputnik

I, II, and III and still not have the ability to go beyond the moon in a lunar probe?

Dr. ZIEGLER. Well, if you talk about electronics and this, of course, is another controversial point—how far the guidance accuracy of instrumentation is—this is just something where we did not have enough facts to make a clear decision.

But, from the brute force approach of just having the thrust available in the rocket, there does not seem to be a problem to get the lunar shot.

Mr. DADDARIO. Do you agree with that?

Mr. GODEL. Yes, sir. It is no problem at all. They have done it.

Mr. MITCHELL. Will the gentleman yield at this point?

Mr. DADDARIO. Just one question. You say they have done it, but the question is, just because they have done Sputnik I, II, and III, does that mean that they can go and have gone beyond the moon?

Mr. GODEL. Well, sir, without interfering with Dr. Ziegler's testimony, I, as you know, represent the Advanced Research Projects Agency and we consider the total information we have and the evaluation of that information to unequivocally confirm the fact that the Soviets did launch Mechta on January 2 as publicly stated.

Did they want it to go in solar orbit between the Sun and Mars, or did they want it to impact the Moon is another problem. But when you define a lunar probe which is getting somewhere in the vicinity of the Moon, my answer has to be they have done it. So, obviously they have the capability.

Mr. DADDARIO. When you say they have done it, on what do you base the fact they have done it?

Mr. FULTON. Let us not jump at the moon like that.

Mr. GODEL. Again within the definition which is within the vicinity of the moon. If you said to impact the moon or circumnavigate the moon, it becomes a judgment thing.

Mr. DADDARIO. I am inclined to that view, except here this morning I hear from Dr. Ziegler this situation could be simulated. If that is a possibility on that particular occasion, then there is another alternative, there is another possibility.

Mr. GODEL. Mr. Congressman, without detriment in any way to what Hans has testified to, I believe it was clear that he indicated within the limits of the reception obtained by the Fort Monmouth facilities working on this thing, it is possible this could have been simulated. There is, however, information, much of it of a highly classified nature, which makes that possibility nonexistent.

It is perhaps fair to say in this context that in respect of Mr. Mallan's article, and in respect of any given fragmentary piece of information, we are somewhat in the position of asking people with limited knowledge, or with restricted knowledge, so draw conclusions which, if they had access to all of the facts that are available to the U.S. Government, they would probably be disinclined to draw.

The conclusion that I am offering here is one that is based upon, I am certain, the totality of the knowledge available to the U.S. Government, from an all-source basis, that is to say—unclassified, classified, at a relatively low level of classification, and very highly classified. And on the basis of this totality of information the conclusion is inescapable that a vehicle launched by the Soviet Union, carrying a transmitter, did traverse to the area of the moon, within four to five thousand kilometers or thereabouts, and then ceased to be heard, by us or the Soviets. What happened after that becomes, sir, a matter of mathematics, not a matter of tracking or a matter of deduction.

The problem is a vehicle having gotten that far would necessarily, as a matter of simple mathematics, have to go into a solar orbit. Whether the accuracy of the figures adduced by the Soviet Union are correct or not, this is anybody's guess because we could not know that they were correct or incorrect.

All that one can conclude, I believe, is that the vehicle, having gotten that far, must have gone into what we call a solar orbit.

It could do nothing else. There is nothing to cause it to break up.

Mr. DADDARIO. Thank you.

Mr. MITCHELL. I might say, Mr. Daddario, very shortly we will go into executive session and these two gentlemen, who are not free at this time to give some of the information you would like, will be here.

Mr. DADDARIO. Yes.

Mr. MITCHELL. Doctor, would you really believe it possible to simulate the lunik signals on 183 as well as 19.9? Do you think both could be simulated?

Dr. ZIEGLER. No; because this would not be an ionospheric frequency. This would be a line-of-sight frequency.

It could not be simulated. If it was heard over a particular area it has to be within sight of the particular area and if it had any antenna directional characteristics, you could determine from what direction it came. Unfortunately we were not able to observe this particular frequency, which, is completely different and could not be simulated.

Mr. MITCHELL. Let me ask you this, concerning Mr. Daddario's line of questioning about the possibility of simulation of the signals you received, were you referring only to the signals you received at Fort Monmouth and you were not concerned with any others?

Dr. ZIEGLER. Only that signal in the area of 20 megacycles which would have that particular characteristic.

Mr. MITCHELL. Could you suggest any reason for the failure of the radar telescope at Jodrell Bank to detect any signals in the region 182.3 even though it was aimed directly at the moon for several hours before and after the announced time of the closest approach?

Dr. ZIEGLER. Well, there are many possibilities. It is hard to say. I understand this particular dish has also side lobes of particular magnitude and has been marching along on other projects, even the American lunar probes, and there can be many small functional defects which can cause such a thing.

Of course, this particular dish has a terrific gain, but it can be upset by a number of malfunctions, and I just do not have any idea what happens here.

Mr. MITCHELL. It might be a defect in the equipment itself.

Dr. ZIEGLER. I just want to mention, in ours we had similar occasions on the part of the American lunar shot at 169 megacycles where we had a rush job to get it finished, and we were not able to protect the converter on this particular day. As a condition we had heavy dew, and we got out of business, due to that moisture, which just prevented us from operating there. Those things happen. They can be prevented as they go along, but doing one rush job after another and getting ready for the next thing we sometimes do not reach that stage of completion that we should.

And things like that could happen there, too. It is hard to say from a remote place.

Mr. VAN PELT. Were you contacted at any time, Doctor, by the author of the article in question?

Dr. ZIEGLER. No, we were not.

Mr. VAN PELT. Was anyone of your staff contacted?

Dr. ZIEGLER. If that would have happened, I would have received knowledge.

Mr. VAN PELT. Thank you.

Mr. WOLF. Mr. Chairman.

Mr. MITCHELL. Mr. Wolf——

Mr. WOLF. I just wondered how many different scientists did you visit with when you were in Moscow for 2 weeks?

Dr. ZIEGLER. I was present at the IGY conference, the International Geophysical Year's special committee, and we were in contact with a tremendous number. I cannot say offhand. I think there were some 50 or 60 Russian scientists available with which we talked at one time or another.

Mr. WOLF. The gentleman, Mr. Mallan, bases his authenticity on the cross-checking he was able to do, between scientists, and each one could or did corroborate the other's position, apparently. I was wondering if your reaction was the same.

There was a very strong opinion which came out here that he was given a "snow job." I would just wonder if your cross-checking produced the same effect as his.

Dr. ZIEGLER. We have hardly ever asked the Russians, "Are you going now to the moon, or are you shooting the next sputnik, or what will the next sputnik contain?" because we have received negative answers on similar things before, and we have more or less given up on asking those particular questions.

Now, as far as scientific questions are concerned, detailed scientific questions, we felt there was a good correlation between the various scientists, and they were cooperative wherever they could be.

But I must say one thing. We had been invited to several institutes where we had great hopes of receiving information; and when we came there we talked to the administrative tops of the institutes, rather than to any people who could give us any information on electronics or space technology.

Mr. WOLF. In other words, you learned about the mechanics of running the station, but not about the work being done by the station.

Dr. ZIEGLER. That is right. Or we went to the Institute of Physics, and in a laboratory or an institute of physics you would expect to see laboratories, but we saw office rooms and files, and we could talk about how they finance their research and how they get the money and how the fiscal year compares with the U.S. fiscal year, but you could not get any detailed scientific results.

Mr. WOLF. I am just wondering if you would like to give any reason why you think he feels he got more information than you did.

Is it his lack of real understanding?

Dr. ZIEGLER. I do not know. That is hard to say. He talked to quite a number of people, obviously.

I may be wrong, but if I valuate the Russian mentality and their overall conditions, personally, I did not think that anybody who goes over and says, "I would like now to discuss the missile and satellite business; let me talk to all of those people," that he gets an honest—that he gets that kind of a job.

Mr. MITCHELL. Let me say we want to have some time with these two gentlemen in executive session. Mr. Godel has not had an opportunity to make any statement. I do not know if you had any prepared statement you wanted to make or if most of your testimony is going to be in executive session.

Mr. GODEL. I might make one if it is useful, Mr. Chairman, not in the nature of a prepared statement, but in the nature of a summary, in open session, if it is desirable, merely to outline the totality of the U.S. facilities that are available for this kind of work, in broad general detail or in broad general scope; and then if you want the detailed physical location and capabilities for them, this might be classified.

If you prefer, I could give the complete picture in executive session, if you wish.

Mr. KARTH. Mr. Chairman, I think there is only one question that should be answered in open session.

Did not someone suggest that ARPA was duplicative and, therefore, should be eliminated?

Mr. WILLIAM GODEL. I believe someone has so suggested; yes.

Mr. KARTH. Why, in your opinion, is it useful, and what does it do that is not duplicated by any other agency?

Mr. GODEL. In one very simple sentence, Mr. Congressman, it provides the Federal Government with a single military space program and not three.

This, I submit, is its unique contribution. In addition to that it has the facilities to do and in the particular case we are talking about here I believe can demonstrate the capability of providing an overview or an oversight of the entire program which will enable us to insure that we have put all of the pieces of a given problem as assigned to us together, and not a part of them.

We believe this to be both economical and efficient; we believe it to be something relatively unique, if I may say so, in military affairs.

Mr. KARTH. Take your time, because this charge, I think, was made before a committee, and you should have your day in court here and try to prove your value or usefulness.

Mr. FULTON. I thought we were looking into the existence of mechta and not ARPA.

Mr. GODEL. We are prepared to go into this in whatever detail the committee desires, of course.

Mr. MITCHELL. It is certainly an important thing, but a little remote from this hearing. Perhaps it should be referred to a subcommittee.

Mr. WOLF. I would like to say the line of questioning Mr. Karth has raised is significant, and while we cannot go into it today, I would hope we can go into it later in detail.

Mr. FULTON. Yes; I think it is a question that should be brought up, in justification of ARPA as well.

Mr. MITCHELL. I agree.

Mr. FULTON. May I have a question?

Mr. MITCHELL. Just a moment, Mr. Fulton, if you will.

Would you rather have Mr. Godel go ahead with his brief general statement?

Mr. FULTON. I would like to question him on a couple of his premises, if you will.

Mr. MITCHELL. Proceed.

Mr. FULTON. My point may not be clear, but it is this: The Russian Government is using for propaganda purposes the mechta as proof they are the first Government to orbit the sun.

They, likewise, claim to be the first people to orbit the earth. I am on the Foreign Affairs Committee, so it is just a little more than a mere academic question to me, and I do not like to see a Russian propaganda position endorsed by scientists, just casually off the back of their hand, with a wave of their hand, without equivocation and as absolute truth.

My point is not whether the mechta or the lunik was launched on January 2, 1959, but that I have to have proof beyond the signals stopping short of the moon. That is, it must be a directional proof, it must be a velocity proof, and it must be an inescapable conclusion that that particular vehicle went beyond the moon and did not settle in the moon's gravitational field.

The Russians themselves will have to provide for me tapes to show signals reaching beyond the moon at a sufficient distance and at a velocity so we can say it was an escape velocity for the moon's gravitational field as well as the earth's gravitational field.

Likewise, I would have to know its ballistic course and the ellipse which, in spite of the second law of Kepler, would take it at escape velocity.

Secondly, the ellipse it entered into as it came in close proximity to the moon's field was not so diversional as to put the original course and velocity out of kilter and spoil its course on an escape velocity.

Next, I would have to know that the vehicle entered into a third ellipse, or was beginning to come into a third ellipse through the sun's gravitational force, and I would have to be shown a tape of some considerable time, from more than two sources tracking on that third ellipse, before I could say it had settled into an ellipse so that in due course by our physical and mathematical calculations, we could say it is a vehicle orbiting the sun.

My comment here, for propaganda purposes, is that I have not seen that kind of evidence, and our tapes are not conclusive on that.

Mr. MITCHELL. Will the gentleman yield?

Mr. FULTON. Yes.

Mr. MITCHELL. I think in view of Mr. Fulton's statement and in view of your previous statement, Mr. Godel, concerning all the answers that would be needed to satisfy Mr. Fulton's question, apparently those questions have been answered in the affirmative as far as ARPA is concerned, according to your previous statement.

Mr. FULTON. No——

Mr. MITCHELL. I may be wrong, but I think Mr. Godel should at this point, as far as he can go in open session, respond to the statement of Mr. Fulton.

48438—59——9

Mr. GODEL. I would be happy to.

Mr. FULTON. Do you have any evidence on tape or other physical evidence or data to show that the vehicle started into an orbit around the sun?

Mr. GODEL. No, sir. As I said earlier the trajectory of this vehicle beyond a point in the general vicinity of 5,000 miles from the moon's surface at that time is for us—and we are firmly convinced, for the Soviet Union—a matter of mathematics.

Mr. FULTON. At what speed was it going according to your last doppler effect?

Mr. GODEL. The speeds——

Mr. FULTON. Was it an escape speed from the earth's gravitational field, as well as the moon's gravitational field?

Mr. GODEL. Yes, sir; our computations would be such, on the basis of the information, independently obtained from any Soviet announcement—our computations would be such as to indicate the vehicle was this: a trajectory in excess of an escape velocity from the earth.

Mr. FULTON. Was it going more than 7 miles an second?

Mr. GODEL. The precise figures I would have to check and I would want to but——

Mr. FULTON. How close were your figures correlated to the point of escape velocity? You must have various figures that showed a pattern or a distribution.

All your data did not show the same result every time you received it?

Mr. GODEL. No, sir.

Mr. FULTON. So what kind of an average did you get?

Mr. GODEL. Insofar as our total tracking is concerned, we were—we placed the vehicle in excess of the 25,000 feet per second velocity throughout the period of the trajectory which would provide it with escape velocity. Obviously, these vehicles do slow down as they go on out. The computations to carry out in the general area of the moon are rather more sparse.

There are a number of indications——

Mr. FULTON. Computations are more sparse, or data?

Mr. GODEL. The data upon which the computation is——

Mr. FULTON. Is very sparse, is it not?

Mr. GODEL. Is very sparse all right, sir.

Mr. FULTON. So the computations are very sparse, too, because they are based on that data?

Mr. GODEL. Except, sir, there are one or two indications which I can refer to later which would suggest that whether or not it was their intent—and this one has to only hypothesize another—the vehicle was not at a position or at a velocity which would permit it to impact the moon or lead it into a spiral so it would ultimately orbit the moon.

Mr. FULTON. The velocity was obviously slowing down on the ellipse it was moving on.

So, under Kepler's second law, how fast was its velocity? How fast was the vehicle losing its velocity?

Mr. GODEL. I do not have the precise figures on hand but I can get the figures that have been adduced for you.

Mr. FULTON. Some of us will not concede that the Russians have a first orbiting of the sun, unless you will likewise concede to us the pellets thrown up into the air by the deep rocket shots that the United States made quite a few years ago likewise orbited the sun.

You see, I am not going to give up.

Mr. GODEL. I have read the pellets of Operation Farside could logically have done this. I would suggest to you from a very limited knowledge of Operation Farside and a more detailed knowledge of this one, that the likelihood of a solar orbit is every bit as great and perhaps greater in the case of mechta, than in the case of the pellets, simply because more data is available.

Mr. FULTON. You then put it on the ground of probability that Operation Farside had less probability than the Russians had, and once you put it on the ground of probability, that is your estimate. That is not conclusive proof to me.

Mr. GODEL. I agree with that, sir.

We are not able to provide definitive proof in terms that you have requested as to the trajectory of the vehicle beyond the vicinity of the moon.

I have indicated that this information, both on the basis of Soviet data and the basis of our own, would be done on the basis of simple mathematical computations, not on the basis of tracking.

Mr. FULTON. Would you, in your judgment, say our tracking tapes were in all respects adequate for a firm judgment on the mechta?

Just answer that "Yes" or "No."

Mr. GODEL. That is impossible, sir.

Mr. FULTON. I want to get your judgment of the data. I want to know how good the tapes are. They were not very good, were they?

Mr. GODEL. The term I found I could not answer was "in all respects."

Mr. FULTON. They were not adequate in all respects.

Mr. GODEL. They were adequate to arrive at the conclusion.

Mr. FULTON. They were inadequate in many respects, looking at the converse of that.

Mr. GODEL. Well——

Mr. FULTON. You would have liked to see them better, wouldn't you?

Mr. GODEL. As I testified before, Mr. Fulton, we are perhaps twice as good——

Mr. FULTON. Answer my question. You would have liked to see them better?

Mr. GODEL. Yes.

Mr. KARTH. Would the gentleman yield?

Mr. FULTON. All right. As long as you said you would like them better, you agree you do not have conclusive proof.

Mr. KARTH. I am also one of those who would not like to give credit to the Russians where credit is not logically due.

I am sure our scientists and these gentlemen here do not want to pat the Russians on the back if and when the Russians did not deserve a pat on the back, probably at the American Government's expense. If I had my "druthers," I would rather see the Russians did not do this successfully. But I feel I will have to form any conclusion on the basis of what these people say, who know what they are talking about, and if they do not, then I do not know whose testimony we can depend upon.

Mr. FULTON. Would you yield?

Mr. MITCHELL. You have the time.

Mr. FULTON. Unless there is basic proof that makes the point a logical point on which to disagree.

Mr. KARTH. It all depends on what kind of proof you want.

What might be perfectly satisfactory to me in the way of proof might not satisfy you.

Mr. FULTON. But let me ask you this.

How about Operation Farside where we put pellets clear out into outer space, so we could never tell whether they returned, and on the last tapes or tracking we had of them, they were going at a high enough velocity into orbit around the sun.

Mr. KARTH. Let us say they did: I do not care about that. We are talking about the lunik and whether it went into orbit.

Mr. VAN PELT. I think this should be for executive discussion.

Mr. MITCHELL. I agree.

Many of the questions phrased by Mr. Fulton are matters that can only be heard in an executive session, but I would like to give Mr. Godel an opportunity to make any further statement at this time concerning the questions and the conversations you have just heard, as far as ARPA's position is concerned.

Mr. GODEL. I would make only one comment in response to Mr. Fulton's most recent inquiry and without presuming to involve myself in the foreign affairs part of our Government, we would certainly in the Defense Department not wish to argue that the Soviets are 40 feet tall.

By the same token, we were not wanting to argue, in connection with Mr. Mallan's article, that they were midgets in the field of space technology.

We think that the conclusions that we have arrived at were logically arrived at, they are arrived at in the legal sense on the basis of all of the evidence that is adduced, and we believe them to be firm, beyond a reasonable doubt.

There is no question but that we would like to and have embarked upon a program, prior to mechta, for the improvement of our U.S. ability to detect, unilaterally, that is, without support of the Soviet Union, to detect, track, and obtain additional information from not only such space vehicles as we might launch, but indeed, as we have previously testified before this committee, to improve our capabilities vis-a-vis—vehicles launched by other countries, including the Soviet Union.

We think this program we have taken is one that will enable us to approach more nearly the 100 percent prima facie case that Mr. Fulton quite correctly suggests we ought to have.

We will necessarily admit that as of today, the best we can say is beyond a reasonable doubt.

Mr. FULTON. Would you do something for me in conclusion? Would you put in the record ARPA's considered judgment as to Operation Farside, on the basis of whether those pellets were the first to orbit the sun, and compare that in probability to operation mechta to lunik of January 2, 1959, of the Russians? That is all, thank you.

Mr. MITCHELL. I do not know whether that would be legitimate within the work of ARPA, Mr. Fulton, so may we qualify that by asking ARPA to make that evaluation if it is within their province? We certainly would not want them to undertake a great deal of work which would take them away from other much more important things, for that project, when they may not have the men available to do it without a great deal of study.

Mr. FULTON. The witness has already given his judgment as to the comparison. I am simply asking he put it on the record.

Mr. QUIGLEY. Will the gentleman yield on that point? I think ARPA has a very definite judgment on mechta. I am not aware that they have a considered judgment on the pellets. I think that is the point the chairman is trying to make. I think there is a real question whether it is within the purposes of this hearing to require them to make such a judgment.

Mr. MITCHELL. Unless Mr. Godel volunteers to furnish that information, if there is no objection, he will not be required to do so.

Mr. GODEL. We have no basis, officially, sir, upon which to make such a judgment now—in terms of the "in fact" occurrences or the "in fact" data that are available concerning Operation Farside.

Mr. MITCHELL. That being so, does the gentleman insist upon it?

Mr. FULTON. I do not insist as to this partcular agency, but I think the committee should establish, first of all, that we are the first in orbiting the sun rather than the Russians, and we do not give up that claim to the Russians so easily.

Mr. MITCHELL. The hour is late. I would like to make the statement before going into executive session, that it is not the position of ARPA or any other agency which has appeared before the committee this week to try to pat the Russians on the back and try to establish them as having a "first."

But the scope and purpose of these hearings was to determine just what information we did have to establish whether or not the Russians had or had not successfully launched a lunar probe. After hearing testimony in both open and executive sessions, my opinion is that they probably did, and that is not for the purpose of giving the Russians a "first". But certainly it would be a dangerous assumption to underestimate their capacity and capability. I think the evidence is abundantly clear that they have that capability, whether the lunik is in orbit around the moon or around the sun.

May we now go into executive session.

Mr. FULTON. May I just comment—I do not disagree with the fact of the Russians having a moon probe in the mechta shot. I am certainly not going to concede that that is in orbit as the first orbit vehicle around the sun unless and until I receive further proof.

Mr. MITCHELL. We will now go into executive session.

(Whereupon at 11:35 a.m. the committee proceeded in executive session.)

25X1

Next 5 Page(s) In Document Exempt

Approved For Release 2004/08/31 : CIA-RDP63T00245R000100290011-7

SOVIET SPACE TECHNOLOGY

THURSDAY, MAY 28, 1959

House of Representatives,
Committee on Science and Astronautics,
Special Subcommittee on Lunik Probe,
Washington, D.C.

The subcommittee met, pursuant to call, at 2:05 p.m., Hon. Victor L. Anfuso (chairman of the subcommittee) presiding.

Mr. Anfuso. The committee will come to order.

Mr. Daigh and Mr. Mallan, would you mind standing up, please? Come forward here.

Give your name and official position to the reporter; first, Mr. Daigh.

Mr. Daigh. My name is Ralph F. Daigh. I am the editorial director and vice president of the Fawcett Publications, New York City.

Mr. Mallan. My name is Lloyd Mallan, author of "The Big Red Lie," a series of articles in True, the man's magazine.

Mr. Anfuso. Will you both please raise your right hand? Do you and each of you solemnly swear that the testimony you will give before this subcommittee in the matters now under consideration will be the truth, the whole truth, and nothing but the truth, so help you God?

Mr. Daigh. I do.

Mr. Mallan. I do.

Mr. Anfuso. Will you please resume your seats.

STATEMENTS OF RALPH F. DAIGH, EDITORIAL DIRECTOR AND VICE PRESIDENT, FAWCETT PUBLICATIONS, NEW YORK CITY; AND LLOYD MALLAN, AUTHOR OF "THE BIG RED LIE"

Mr. Anfuso. The question which this subcommittee is going to consider is to try to determine by a series of hearings whether or not the facts concerning the lunar probe as already testified to by Government witnesses and other witnesses are correct as already testified to, and whether or not additional facts can be brought out to clarify what has already been testified to for this committee and the entire Congress.

Of course, all of these questions and answers will be concerned with the lunik probe made by the Soviet Union as it involves international security and peace and the very serious charges made by Mr. Lloyd Mallan in True magazine that lunik was a hoax.

Mr. Mallan is going to be given every opportunity to state his case, and we are then going to call Government witnesses which will either sustain or disprove his conclusions.

I am now going to outline the procedure which this committee will follow.

Mr. Ralph Daigh will make a short statement and introduce Mr. Mallan. Mr. Mallan will then make his statement without interruptions, and then he will be subject to questioning by the chairman and the other members of the committee in the proper order.

Mr. Daigh, you may proceed with your prepared statement.

Mr. DAIGH. Thank you. Gentlemen, it is by my considered responsibility that a story titled "The Big Red Lie" is appearing in the May, June, and July issues of True, the man's magazine. In this story it is our firm assertion that Russia is miles and years behind the United States in all aspects of technology.

Of specific newsworthy interest is the story's statement that lunik, the alleged moon rocket, said by the Russians to have been fired into oribit around the sun January 2–4, does not exist.

We will offer solid proof that lunik does not exist.

In addition to proof of lunik's nonexistence, I will now relate some brief facts which tell their own story as to why the Russians claim, and are able to claim, lunik does exist.

Item 1: A few hours after lunik was announced as launched by the Russians, President Eisenhower publicly congratulated the Soviets on their accomplishment.

Item 2: Recently when queried in a press conference about his speedy commendation of the Russian achievement, President Eisenhower expressed a hope that his congratulations did not constitute an aid to Russian propaganda.

It is a regrettable fact that President Eisenhower was a victim of a Soviet hoax. He and his advisers fell headlong into a Soviet propaganda trap. Lunik is another in the long chain of lying claims made by the Russians, to which our press and official statements have tended to give confirmation.

Item 3: In September or October of 1958 our space program officials announced we would attempt a lunar probe in January of 1959.

Item 4: In December of 1958 we successfully launched a satellite rocket, Project Score, that broadcast a Christmas message to the world in President Eisenhower's own voice, thus winning a short-lived and temporary propaganda victory that was embarrassing to the Russians.

Item 5: In the first week of January 1959, Anastas Mikoyan flew into this country on the heels of the Soviet's claim to have first put lunik into orbit around the sun, on the second day of the very month we had previously announced as the date for our own launching.

Mikoyan thus gets, for Russia, the honor of being the first man launched into space by a rocket—a nonexistent rocket at that. Two more Russian "firsts," as valid as her absurd claims to have invented television and penicillin.

Mr. Mallan is appearing before this committee, as am I, voluntarily. We offered to give our evidence in closed hearing if security purposes were better served.

However, we welcome the open hearing because it is our conviction that a fear phobia is being built up in the free world as concerns Russia. It is only by putting our self-proclaimed enemy in true perspective publicly, as an inferior nation and an inferior system of government, that the free world can develop her strengths, in harmony and in confidence and without panic.

I cannot fail to mention in passing that not only have our responsible leaders been unwitting tools of the phony Soviet claim to greatness but this pro-Russian propaganda has also reached our schoolchildren, with the result that many of them picture Russia as a country whose technological achievements exceed our own. Thus, they are absorbing a propaganda poison that tends to force acceptance of communism as a superior form of government.

At a previous congressional hearing on this subject, a well-meaning Congressman asserted that someone, Mr. Mallan, author of "The Big Red Lie," had been given a "snow job."

We agree that a "snow job" is being done, but not on one science writer, but rather an unending blizzard of a "snow job" is being poured on the Congressman and the whole free world.

It is my opinion that this country must continue to spend all funds necessary to maintain superiority in space and astronautics. I am not concerned with the politics of the situation, or interdepartmental rivalries and battles for appropriations and control. My only concern is that the people of a democracy are entitled to truth, and, given the truth, they will continue to find their way.

Mr. Mallan, author of "The Big Red Lie," will present evidence to prove lunik nonexistent.

Mr. Mallan spent several months in Russia in 1958, interviewing 38 of Russia's leading scientists.

He went into Russia at our suggestion to write books and magazine stories intended to warn our people of the tremendous competitive technological achievements supposedly made by the Russians, an impression that we were under at that time.

As he went from one top scientist in Russia to another, inspected their scientific machinery, discussed theory and fact with these scientists, it grew on him, slowly but surely, that Russia was not ahead of us, but miles and years behind us. His conclusions were based in part on similar interviews he had in this country with our top scientists, and after seeing their machinery.

I would submit, as an editor, that a digging, science reporter cannot be fooled by 38 scientists who are proudly showing off Russia's best so that a Western reporter may write favorably about the Soviet.

These men are scientists, not actors. As scientists, they have a preoccupation with fact, and it is inconceivable that scientists would lie about their lifework and specialties, whether it be astronomy, space tracking, or missile guidance.

Or, at the very least, they could not lie convincingly to a trained science reporter, not 38 scientists.

A word about Mr. Mallan. He is the author of five books. He attended Carnegie Tech. He has literally spent years observing and writing about our own space program and is intimately familiar with astronomy, astronautics, aeronautics, astrophysics, missile tracking, space medicine, radio. He has Defense Department press clearance to visit our own missile centers and is the best man I know to make trustworthy conclusions concerning anyone's space program.

Dr. Fred Singer, who proposed the first feasible earth satellite, says of him:

Mr. Mallan is a science writer of great talent. He has produced important and informative books and articles dealing with American work in astronautics and all related sciences.

James J. Harford, executive secretary of the American Rocket Society says:

Lloyd Mallan, a member of the Rocket Society, is a competent writer who has written about the American rocket and space flight program copiously.

Richard E. Shope, the Rockefeller Institute for Medical Research:

Mr. Lloyd Mallan is a science writer of great talent, whose interests and writings lie in the field of astronautics, astrophysics, guidance and navigation, and particularly space medicine.

Andrew G. Haley, president of the International Astronautical Federation:

Mr. Mallan is highly regarded by me and by professional men in the United States as a science writer.

These are Mr. Mallan's credentials. I am sure you will lend credence to what he has to say.

Mr. Anfuso. Mr. Mallan, you may proceed.

Mr. Mallan. Gentlemen, when I appeared before your full committee several weeks ago, I began to give testimony to the effect that lunik, the alleged Russian moon probe, was a hoax. I appreciate the opportunity to continue my testimony.

Incidentally, at a previous committee hearing, it was suggested that my story of Russia's technological weakness might give aid to our Communist enemy. I want to state now for the record under oath that I am not a Communist, have never been a Communist, and despise all aspects of Communist ideology.

When Dr. Wernher von Braun testified in January of this year before the Senate Preparedness Committee and your committee on the existence of lunik, he made three grevious errors of judgment and fact.

Item 1. Dr. von Braun testified that the Palomar Observatory in California "actually photographed or at least physically observed lunik." Dr. H. W. Babcock, assistant director of the Palomar Observatory told me: "Dr. von Braun's information must have come from a misleading source. There was no photograph or observation made here that could be remotely identified with the Russian lunik." I submit this letter of confirmation from Dr. Babcock in evidence.

Mr. Anfuso. It will be received in evidence.

(The document referred to is as follows:)

Mount Wilson and Palomar Observatories,
Carnegie Institution of Washington,
California Institute of Technology,
Pasadena, Calif., May 18, 1959.

Mr. Larry Eisinger,
Editor in Chief, Fawcett Books,
New York, N.Y.

Dear Mr. Eisinger: The Mount Wilson and Palomar Observatories do not have any photographic evidence either for or against the reality of the Russian lunik last January. Since we have nothing to contribute, we specifically request that in the articles and books appearing under your control, no reference be made to the Mount Wilson and Palomar Observatories in connection with the Russian lunik.

Very truly yours,

H. W. BABCOCK.

Mr. MALLAN. Item 2: As another piece of proof, Dr. von Braun referred to the fluorescent sodium cloud that was allegedly released from the rocket out in space. This cloud, said he, "was actually photographed by at least one station in northern England."

I refer you to page 244 of the record of those same joint hearings. On that page is a statement translated from Pravda, official Communist Party news organ. The statement says that the sodium flare could be seen only from "parts of Middle Asia, the Caucasus, the Near East, Africa, and Asia." In other words, from the Southern Hemisphere. Where is England? The Northern Hemisphere. Hence, even if there was a sodium cloud at the time and place claimed by Russia, nobody in England could have seen it. The photograph mentioned by Dr. von Braun as proof No. 2 was taken in Scotland, and has since been completely discredited.

Item 3: Dr. von Braun made much of the fact that certain weak, sporadic radio signals were received by the Jet Propulsion Laboratory's tracking station at Goldstone Dry Lake, Calif. These signals, he said, came from the point in space where the Russians claimed lunik to be at that time. The signals were received at one of the frequencies Russia had announced.

This is a sad little piece of "proof." Most of the important radio tracking stations in the free world had been alerted several days previously to be on the watch for a Russian moon probe. Everyone, supposedly, was listening. Only the Goldstone station heard a thing on 183.6 megacycles.

In England, the giant radiotelescope at Jodrell Bank was sharply tuned to the same frequency as Goldstone, and scanned the sky January 3 and January 4. This was the same telescope that later followed the American moon probe out to nearly half a million miles. It is nine times as sensitive as the Goldstone equipment. Yet it found no trace of lunik.

In Japan, tracking scientists who had received signals from all other space vehicles heard nothing from lunik. "We cannot acknowledge the existence of the rocket on the basis of our own findings," they stated.

At Boulder, Colo., a scientist of the National Bureau of Standards, Central Radio Propagation Laboratory, Dr. Gordon Little, told me:

We did not hear anything. There was no success whatever.

The American Radio Relay League, with a worldwide association of 90,000 radio amateurs, had collected hundreds of reports from its members on every other space vehicle. Lunik? Not a trace.

What it all boiled down to was this: Many skilled technicians and scientists and many fine instruments had hunted for lunik. Only one station is offered in proof. Yet that lone station, that tiny minority, convinced some people that lunik exists.

It was an illogical and ridiculous situation. It was as though we were told that, some time today, Marilyn Monroe will walk into this committee room, complete with sodium flare. Fifty men sit here all day, watching for her. At the end of the day, we compare notes. It turns out that nobody saw anything—except one man, who thinks maybe he caught a fleeting glimpse of her.

What would the consensus be: was she here or wasn't she?

By any kind of logic, we must conclude that she wasn't—that the lone man was in some way deluded or mistaken. If we are going to be rational, we must say the same of the Goldstone tracking reports: those signals were not from lunik.

I talked with the top tracking technicians at the Goldstone station itself. Pinned down, they were forced to admit that they could not be certain about the signals they had heard.

"There is no way of being absolutely certain. We are fairly sure," Walter Larkin, director of Goldstone, told me.

William C. Tilkington, signal characteristics expert of Goldstone, sounded even less sure. He told me:

> The pulse rates of the signals we received were very different from the ones broadcast by Moscow Radio.

In addition, Dr. John D. Kraus, the director of the Ohio State Radio Observatory, in response to a direct question, says:

> It certainly appears highly questionable that those (Goldstone) signals were from a space vehicle. From that kind of data, I wouldn't be convinced.

These significant statements certainly do not support Dr. Pickering's testimony. Neither do they support Dr. Wernher von Braun's offering of the alleged Goldstone tracking as evidence of lutnik's existence.

In previous testimony before this committee Dr. Pickering described these signals as "very weak." Well, Dr. Pickering, why did not any other powerful and better equipped tracking station pick up these very weak signals? Was Goldstone uniquely lucky?

Also in previous testimony Col. Linscott Hall, deputy director of Project Spacetrack, admitted the data was "extremely scanty." Why does Colonel Hall rest his case on "very weak" signals and "extremely scanty" data?

And finally, in previous testimony, Dr. Fred Singer, University of Maryland physics professor, not a radio astronomer, cites "overwhelming evidence" to support the lunik claim. I ask you, Dr. Singer, are very weak signals and extremely scanty data in any way related to overwhelming evidence?

I submit that this overwhelming evidence is on the other side of the fence, that it rests with the remaining huge majority of the free world tracking stations which never got a peep or heard a beep from this alleged lunik.

Just what were those mysterious signals, then?

They could have come from several sources, but not from lunik. The four weak, brief noises heard over a 5-hour period at Goldstone might more reasonably, according to Dr. Kraus, have come from airplanes, from celestial bodies, radio or television stations on earth.

I am sure that what I have said so far certainly casts more than a reasonable doubt as to the existence of lunik. But I have further proof that lunik is a hoax. I had just returned from a scientific reporting investigation in Russia. While in Russia, I'd made a painstakingly careful study of Russian progress in space technology.

In support of arguments which I present here, let me say that I believe my investigation was the most thorough, first-hand study of Russian space science ever made by a Western reporter.

It's certain that I talked with more top Soviet scientists than any other free world journalist. Moscow correspondents were amazed when I showed them the list of scientists I had talked to.

Why had no other reporter talked to these men?

It is my belief that no one else had ever tried to. Or, if anyone had, he was unable to overcome the obstacles I found.

It is a fact, too, that unless a reporter has a wide range of knowledge in pure and applied sciences, he would learn nothing of Russian accomplishments from talking to these men.

Given such a knowledge of science, I guarantee that any other competent, persistent reporter would come out of Russia with a much different picture of Russian missile progress than that painted by the propagandists in the Kremlin.

Suppose a propagandist told you: "I have a horse that can climb trees, swing by its tail and sing the Star-Spangled Banner." You'd laugh. You'd know that a horse had no claws with which to climb trees, that its tail won't support its weight, and that its vocal cords are unsuited to singing. You would know that the story must be a phony.

That was precisely my reaction to the lunik story. I knew that Russian science was not capable of producing anything so clever or so sophisticated as this alleged moon rocket. I knew, in particular, that the Russians lack seven vital devices—seven devices that are absolutely essential for any such accurate, long-distance space probe:

One, the Russians have no guidance system accurate enough to bring a rocket within 4,650 miles of the moon as they claim.

Two, they have no advanced electronic miniaturization.

Three, they have no miniature electronic computer.

Four, they have no atomic clock.

Five, they have no efficient optical tracking equipment.

Six, they have no efficient radio tracking equipment.

Seven, they have no responsive, miniature servomechanisms.

A nation must have every single one of these devices before that nation can hope to launch a successful sun-orbiting rocket or an ICBM. Russia lacks all seven.

Thus, through evidence I'd gathered on the other side of the Iron Curtain, I knew that lunik could not and did not exist.

So, gentlemen, in stating that lunik does not and never did exist, I've drawn on two main areas of evidence:

First, the fact that no one on either side of the Iron Curtain has submitted to us any proof—or even any reasonably good indication—that such a rocket was either built or successfully launched.

Second, the fact that Russia does not have the scientific or engineering ability to build such a rocket.

Gentlemen, in addition to the above I have some new information which I can only characterize as startling. I have here with me the June 1959 issue of "C.Q.," the American Radio Amateur's Journal, which I will place in the record.

(The information requested is as follows:)

Mr. ANFUSO. Supposing, Mr. Mallan, we just place in the record the article with reference to what you are saying.

Mr. MALLAN. Fine. On page 38 is a story with the following headline: "Russian Hams Pick Up Lunik Signals Easily." As you know, hams are amateur radio operators. This story in C.Q. is based entirely on reports published in the January issue of Radio, the official organ of Russian radio hams. The story quotes profusely from the Russian source to the effect that signals were allegedly received from lunik all over Soviet Russia by dozens of Russian hams.

Radio contact, according to the Russian publication, was sustained on all three frequencies. Signals were reported as loud and clear and contact was maintained for periods of an hour or longer. Contact was maintained to a reported distance of 280,000 miles and "signals were still very loud."

Gentlemen, I have seen Russian ham radio equipment. It is on a par with that we had 20 years ago when I was a ham. And while I cannot testify to the quality of the individual amateur stations, allegedly receiving lunik's signal from Stalingrad to Chita to Khabarovsk, I do know that our professional equipment, including Jodrell Bank's and every free world tracking station, is better than that of the Russian amateurs, or anything the Soviets possess.

How can you explain lunik except as a hoax?

Thank you very much.

RUSSIAN HAMS PICK UP LUNIK SIGNALS EASILY

Joseph Zelle, W8FAZ, technical staff, WERE AM-FM-TV, Cleveland, Ohio

When the Ruissians fired their lunik last January 3, the radio signals were apparently picked up in Russia without any trouble. Even radio amateurs beyond the Iron Curtain at once tuned in the three frequencies given out by Moscow Radio. This success was in sharp contrast to the Western World, where no tracking station was able to identify the lunik signals positively.

One of the first to receive the radio signals in Moscow from the spacecraft was N. Kazanskey, UA3AF. For 10 minutes he picked up telegraph transmissions at 0040 MST on two frequencies, 19.993 and 19.997 megacycles with a strength of S6–7.

Several others in Moscow heard the voice of the rocket. R. Hauckman, UA3CH, heard it at 0830 MST. A little later B. Mashkov, a young operator at the Pioneers' Home station heard it. V. Kozlov and T. Shcheglov, radio operators of the Central Radio Club, UA3KAA, even succeeded in recording the signals on tape.

In Stalingrad, U. Byertyayev, UJ8AG, copied the signal for 10 minutes at 0930 MST. In Gomyel, UC2KAB monitored the transmitters clearly from 1145 to 1155 MST, when the lunik was some 125,000 miles away (200,000 km.). UB5KCA reported 19.995 megacycles S6 at about 1500 hours MST, when the Soviet cosmic rocket following its charted course, had passed more than half the distance to the moon.

Ten radio amateurs in Chita, standing radio watch, likewise heard the signals. Grigoryev, an instructor of their DOSAAF Radio Club, recorded the signals on tape. At the collective radio station, UA1KMC of the Borovichi Radio Club, operators made a fix on the 19.997 megacycle frequency at 0813 MST, January 3. Estonian radio hams were successful too, as reported by radio station UP2KAA of Tallin. In Kursk, UA3WZ, O. Kolozin was successful in tuning in the rocket signals several times.

Radio station UO5KPM of Kuibishyev reported many Moldavian radio amateurs had heard the cosmic rocket radio signals. Among these hams was veteran shortwaver, A. Kamalyagin, UA4IF.

Later, signals were reported from Batumi. At 2036 MST, January 3, the scientific observation station located in the eastern part of the Soviet Union picked up the signals early. Unlike the unsuccessful early attempt at Jodrell Bank, Batumi reported hearing signals considerably down below the horizon on the following repickup of rocket signals. Soviet radio amateurs in the area stood radio watch along with the station. From 2100 to 2130 hours, radio amateur G. Sidoryenko, UF6PB, heard the signals very loudly.

On January 4 at 0543 by Moscow time, the Soviet rocket reached the closest point to the moon, according to Soviet scientists. At that time, M. Livonsky, UA1BG, of Leningrad, monitored the rocket radio for about an hour, with signal strength reported at RST 578.

The most surprising report came from Lower Tagil, by the operator of radio station UA9KCC of Smolin. The signal was picked up at 1557 MST on January 4. At 1725 the signals were still very loud. At that time, the Soviet rocket was located about 280,000 miles (450,000 km.) from the center of the earth.

All these reports from radio hams were of a preliminary nature. They were included in the January issue of Radio the official, organ of Russian amateurs, DOSAAF. Just how many of these reports are questionable is hard to determine.

Conversely, no professional tracking and observing stations in the Western World reported hearing the lunik signals. No United Kingdom tracking stations had any success, including Prof. A. C. B. Lovell's huge radiotelescope at Jodrell Bank. In America, three nonprofessional listening stations compared signals recorded on tape, but all three were different. Moreover, none of the three taped sounds agreed with the lunik sounds officially broadcast by Radio Moscow on January 3.

The writer himself monitored the three listed frequencies, 19,993, 19.995, and 19.997 megacycles and heard signals. However, these were highly questionable as they were far too loud and too commercial in sound to be considered to have originated 68,000 miles away.

One difficulty was the presence of WWV on 20.0 mc which came through fairly strong even late in the wintry evening. This situation excluded the 19.997 mc (probably the strongest signal) as well as 19.995 mc in its weak condition. The chances of 19.993 mc signal being continuosly on the air was also questioned. The transmissions may have been occurring at widely scattered periods, for short intervals, which might also account for missing the signals.

The Russians have since made a request for additional reports and recordings of lunik signals. They appealed to professional and amateur stations throughout the world.

Meanwhile concerted efforts were being made in the United States to determine the cause of our odd failure to pick up lunik in outer space.

Mr. Anfuso. Mr. Mallan, I understand you have another statement from some university. Do you have it with you?

Mr. Mallan. I have an unsolicited letter from Trinity University from Dr. Paul Seabase, who is chairman of the physics department. Shall I read it aloud?

Mr. Anfuso. No, I don't think so.

Mr. Mallan. Shall I include it in the record?

Mr. Anfuso. Give us the date of that letter.

Mr. Mallan. The date of the letter is April 28, 1959.

Mr. Anfuso. Does that letter support your contention?

Mr. Mallan. Yes, it does.

Mr. Osmers. There being many "Trinities," where is this one located?

Mr. Mallan. This one is located at San Antonio, Tex.

Mr. Anfuso. For the record, I might say before I ask you some questions, Mr. Mallan, that Marilyn Monroe was not here.

Mr. Mallan. Thank you.

Mr. Anfuso. Mr. Mallan, doesn't the weight of the sputniks, especially Sputnik III, prove that the Soviets have enough thrust to send a rocket to the moon?

Mr. MALLAN. I am not sure of what the weight of Sputnik III is except for the figures that the Russians have released.

Mr. ANFUSO. What was that figure?

Mr. MALLAN. Approximately a ton and a half.

Mr. ANFUSO. Assuming that Sputnik III was a ton and a half, doesn't that prove that the Soviets have enough thrust to send a rocket to the moon?

Mr. MALLAN. But I can't assume that Sputnik III weighs a ton and a half.

Mr. ANFUSO. You mean there is no evidence that it did weigh a ton and a half?

Mr. MALLAN. That is correct.

Mr. ANFUSO. Is it also your contention that no one has come forth with proof that it did weigh a ton and a half?

Mr. MALLAN. There is no proof so far as I know.

Mr. ANFUSO. With such a propulsive capability—assuming they had such a propulsive capability—couldn't they send a complete ICBM system into orbit or to the moon?

Mr. MALLAN. No, sir.

Mr. ANFUSO. Why not?

Mr. MALLAN. Because thrust alone is not the thing that makes a missile or a space probe.

Mr. ANFUSO. What does make a missile or a space probe?

Mr. MALLAN. The most important part of it is the guidance system.

Mr. ANFUSO. You say they lack a proper guidance system?

Mr. MALLAN. I do.

Mr. ANFUSO. Do you have some proof of that statement to demonstrate to this committee?

Mr. MALLAN. Well, I have the proof of having seen their equipment, and I can also state that the Russians claimed that lunik passed within 4,650 miles of the moon.

In terms of outer space distances, that is a very near miss. It is a superb piece of celestial marksmanship. To achieve that kind of accuracy you need an extremely precise, reliable system for starting your rocket in the right direction and holding it on course.

There are two kinds of guidance systems that can give you this kind of accuracy. One is radio control from the ground. The other is inertial navigation. They could not have used ground control radio because if they had, the signals would have been picked up by free world tracking stations. The type of equipment used in remote radio control generates very powerful signals. These radio signals would have been heard all over the world by our monitor stations just as surely as we hear a gunshot fired in the back of this room.

The Russians could not have used inertial guidance either. The reason is simply that they have none to use. Inertial guidance depends on a complex set of miniature instruments which sense and control the rocket's motion in flight.

To build such a system, you need miniaturization, high precision engineering, and a very good but very small computer. These things are beyond the Russian's capabilities. I became convinced of this, convinced that the Russians couldn't build the kind of inertial guidance system needed, simply by studying Russian engineering. In addition, I asked Russian scientists about their experiments in inertial guidance. Many scientists I asked had never heard of inertial guidance. But the Soviet Academy of Sciences, the fountainhead of all Soviet scientific activity, I talked with Profs. Yuri Pobedonetsev, Kirel Stanyukovich, both key scientists in the Soviet space program. I asked point blank whether Russia had developed an inertial guidance system. Russia would like us to think she is far ahead of us in the conquest of space. I knew that if these men chose to lie to me they would lie in the direction of exaggerating Soviet progress and Soviet cleverness.

But they didn't lie. This was Stanyukovich's answer to my question:

> Some work in this field has been done, but it is all theoretical so far.

Mr. ANFUSO. So you say that they could not have been as accurate as they claimed?

Mr. MALLAN. That is correct.

Mr. ANFUSO. Are you also supported in that contention by this university professor who wrote to you?

Mr. MALLAN. Yes; I am.

Mr. ANFUSO. Will you identify him again for the record? He is chairman of the physics department of what university?

Mr. MALLAN. Of Trinity University, at San Antonio, Tex.

(The documents from Mr. Seabase are as follows:)

TRINITY UNIVERSITY,
SAN ANTONIO, TEX., *April 29, 1959.*

FAWCETT PUBLICATIONS, INC.
Greenwich, Conn.

GENTLEMEN: I would appreciate it if you would be so kind to transmit the attached letter to Mr. L. Mallan, whose address I do not know. It is a comment in reference to his article "The Big Red Lie" in True magazine. I would like to get in touch with him because he expressed in his article exactly the opinion I formed when the alleged launching of lunik was publicized.

Yours truly,

PAUL L. SEABASE,
Chairman, Physics Department.

TRINITY UNIVERSITY,
San Antonio, Tex., April 28, 1959.

DEAR MR. MALLAN: I have read your article in True magazine about the "Big Red Lie." I read it and reread it carefully, because this your article expresses just exactly the opinion I formed when the first propaganda blasts came from Moscow.

I am the leader of the San Antonio Moonwatch team (089–029–098), thus somehow professionally involved. On the other hand I am a refugee from Hungary after World War II. I have been as a member of the Hungarian R.A.F. during the war in occupied Soviet territory. I have seen and as a mechanical engineer I could evaluate the technical and scientific potential of the Soviet as far as it stood at that time. They may have caught up since 1943, but I also know as an educator that an educational gap which widened and deepened from 1917 until before the World War II cannot be closed with a couple of 5- and 7-year plans. I believe that their backlog in training and education may be leveling off in a couple of decades ahead if they do it properly.

The controversial announcements about the pass of lunik by the moon arose my first suspects. I have on tape the original Moscow figures which said that lunik passed 4,000 miles from the moon at 26,000-mile-per-hour velocity. These two figures are so tremendously controversial that I did express my opinion right then, before my analytical mechanics class, that there is not any lunik existing.

Then came the so-called photographic evidences. As a trained amateur photographer I do know how such pictures can be faked. The skin temperature for lunik they quoted in comparison with the Pioneer's is again a clever mixup of units because 60° C. is about 150° F. Numerically there is quite a difference between the two degrees. But considering the units the temperatures are identical. The alleged sodium-cloud picture of the Scottish amateur's looks to me rather than a lenticular cloud in the upper atmosphere. It is doubtful if an amateur would be able to take an excellent picture from a phenomenon where astronomers with professional equipment failed. The same stands for the radio signals.

I have been interviewed by a reporter of our local newspapers about your article and I was very pleased to give my opinion fully confirming yours.

Claims of the Soviet, through the propaganda trumpet, may make the layman believe. As a scientist I like to double check from every possible corner. When months ago under the influence of the announcements, my opinion was rebuffed or ridiculed locally, I did hope, someday someone will come up with the same findings and clear the haze.

With best regards yours,

PAUL S. SEABASE,
Chairman, Physics Department.

Mr. ANFUSO. Mr. Mallan, isn't the state of Soviet space technology in other respects—guidance and miniaturization, et cetera—really irrelevant to the question of whether the Soviets could have shot a rocket past the moon on January 2, 1959?

Mr. MALLAN. No, sir; they are not irrelevant.

Mr. ANFUSO. Will you explain that for us?

Mr. MALLAN. In order even to shoot a heavy weight into space that close to the moon you require a very compact set of instruments. However, I think I could sum it up more graphically by stating that if they had one-five-hundredths of a second's error in the total timing of the mechanism involved and the electronic devices, they would have missed the moon by at least 100,000 miles and probably much farther than that.

Mr. ANFUSO. I think you said this in your previous testimony, that we have perfected miniaturized components; isn't that correct?

Mr. MALLAN. We have not only perfected miniaturization of components, but we are now working in a field called microminiaturization, which makes use of sizes comparable to pinheads.

Mr. ANFUSO. How are the sounds from the satellite that we now have in orbit? Are they clear and significant?

Mr. MALLAN. Our own satellites?

Mr. ANFUSO. Yes.

Mr. MALLAN. Yes; there has rarely, if ever, been any trouble in receiving signals and analyzing data from our satellites.

Mr. ANFUSO. Are they picked up by hams as well?

Mr. MALLAN. Yes; they are.

Mr. ANFUSO. What was the size of that satellite?

Mr. MALLAN. Well, we have had a variety of sizes, starting with about 3½ inches in diameter and weighing 6 pounds, I believe. Our heaviest one was around 31 pounds. The solar probe, or the lunar-solar probe, Pioneer IV, I believe weighed 18 pounds.

Mr. ANFUSO. We can hear the sounds from each one of those very distinctly?

Mr. MALLAN. Yes, sir.

Mr. ANFUSO. How about the so-called lunik? Do we hear those sounds today?

Mr. MALLAN. No. As a matter of fact, only one tracking station in the whole free world heard signals from the lunik on one frequency and another frequency. In other words, two stations as a total claimed to have heard these signals, neither of which has substantiating scientific proof of this.

Mr. ANFUSO. Which of those two stations?

Mr. MALLAN. The Goldstone tracking station of the Jet Propulsion Laboratory, and the tracking station at the Stanford Research Institute at Menlo Park, Calif.

Mr. ANFUSO. Is it the claim of those two stations that they are still hearing sounds?

Mr. MALLAN. Oh, no; Goldstone heard signals only sporadically over a period of 5 hours approximately. They heard four bursts of signals, the longest of which was 3 minutes in duration. Standard—I believe they sent a telegram that is in the record—listed the frequencies and the types of equipment they used to hear lunik's signals, and their frequencies were on the ham frequencies or near to it.

Mr. ANFUSO. In other words, they were only heard for the first few hours and have not been heard since, is that correct?

Mr. MALLAN. That is correct.

Mr. ANFUSO. Do you think that those sounds which were heard by Goldstone could have been simulated, or is there conclusive proof that they came from lunik?

Mr. MALLAN. It is impossible for those signals to establish conclusive proof.

Mr. ANFUSO. Could they have come from some other source?

Mr. MALLAN. They certainly could have come from some other source.

Mr. ANFUSO. What, for instance?

Mr. MALLAN. I can show you—as a matter of fact, I have a chart that I believe is very graphic. I suppose I should take it up where everybody can see it.

This is a graph of the Goldstone radio telescope. This, of course, is a much simplified graphic description, but it is based on the aeronautical chart, radio beacon chart, for the Mojave Desert area where Goldstone Dry Lake is located. In this area there are two main airways over which traffic is very heavy. As a matter of fact, the Federal Aeronautics Administration told me that an average 24-hour period for instrument flights alone runs from 65 to 75 flights. These are only instrument flights, and here we are only indicating the omnirange stations which send very high frequency radio waves out in all directions.

There are many other stations, including stations of the Air Defense Command, which are also on both VHF and UHF frequencies. There are also privately owned stations owned by airline companies themselves.

Many of these stations are code, not voice stations, and many of them have carriers that would have a modulation on it similar to the type described to me by some of the crew at Goldstone. At any rate, there are three possible ways that these terrestial signals could have been received by the Goldstone antenna and given the impression that the signal source was moving away from the moon, because at Goldstone they base—or rather Dr. Pickering bases his entire attitude on the fact that these were alleged lunik signals—he bases this on the fact that the angle of the signal source widened in relationship to the moon, over 6 degrees in a period of 5 hours.

However, if an airplane, just one airplane, came along when the antenna was scanning in this direction toward the moon and an airplane was flying across the standard airway, it could have picked up a beep hermonic signal—I will try not to be too technical.

Mr. ANFUSO. Please do not.

(The graph referred to follows on p. 151.)

Mr. MALLAN. It could have picked up a signal from a radio-emitting source on earth which would bounce against the airplane's metal body and reflect right down into one of the minor lobes of this big antenna.

Then, as the moon kept moving, another airplane coming from another direction but still crossing where these beams go out in all directions could have also bounced another signal off. And the same thing is true here. There are many other possibilities. A channel 8 radio television station—hundreds of miles away, in fact—could have bounced a signal off passing airplanes at high altitudes down to the Goldstone dish, which was tuned to channel 8.

Even police car radios operate at frequencies which have harmonics in the band at which the Goldstone antenna was tuned.

Mr. ANFUSO. Is it not the claim of Goldstone that they heard these sounds for a steady period of at least 4 hours?

Mr. MALLAN. No; the director of Goldstone, Walter Larkin, told me—and I have a verbatim transcript of it—that the signals first on one occasion he said were medium to zero, and another he said from medium weak to nonexistent, and that they heard four bursts of them in the space of 5 hours. The longest burst of signals lasted only 3 minutes, according to the director of the station.

Mr. ANFUSO. Let me interrupt you again. These bursts were heard at the specific time that the Russians had already stated in advance they would be heard; is that right?

Mr. MALLAN. No. The Russians had announced that at this point the alleged lunik—if you will permit me—was beyond the moon, had just passed beyond the moon. Goldstone was scanning the sky on both sides of the moon trying to find this.

Mr. ANFUSO. Before the shot called lunik, did they not make some announcement to the press that on such and such a date they would make this probe?

Mr. MALLAN. No; they gave no advance announcement.

Mr. ANFUSO. Then I was mistaken.

Mr. MALLAN. They announced it after the fact.

Mr. ANFUSO. Where was Jupiter in relation to the moon at the time of lunik's closest approach to the moon?

Mr. MALLAN. Jupiter is another possibility. Jupiter has a continuous—it transmits signals, to put it simply—on a continuous band of wavelengths all the way from 18 megacycles up to a few thousand, or 6,000 megacycles, in fact. There is a possibility since these signal characteristics on the graph reductor Dr. Dryden sent me was proof that these signals were from the alleged lunik, the signal characteristics follow the pattern of bursts of signals that are emitted from Jupiter in the 18- to 30-megacycle range. They emit short bursts for

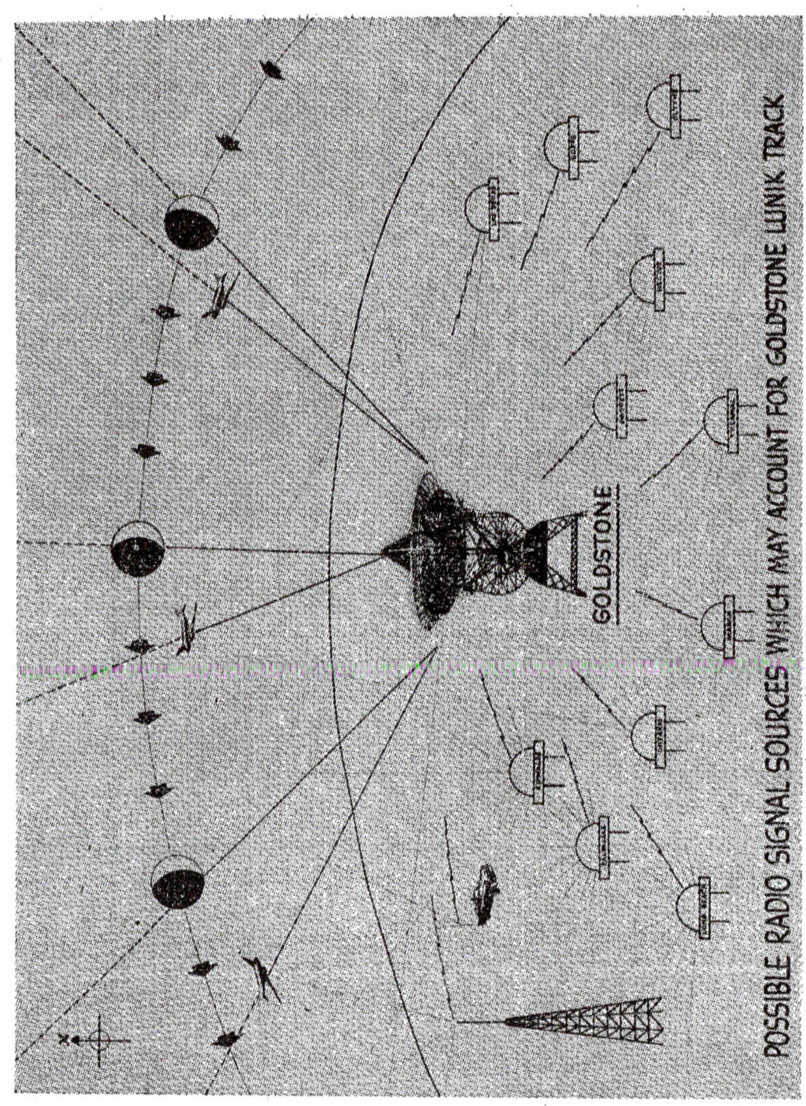
GOLDSTONE
POSSIBLE RADIO SIGNAL SOURCES WHICH MAY ACCOUNT FOR GOLDSTONE LUNIK TRACK

a few minutes and then there is a long silence for an hour or two and some more bursts. That is not true of the higher frequencies on which Jupiter transmits.

I understand Dr. Pickering gave testimony that Jupiter's signals were very easily recognizable since they sounded like lightning flash, if a lightning flash can make a noise. It is the thunder afterward.

But, at any rate, this is not true. This is only one type of signal from Jupiter. There are three types.

Mr. ANFUSO. Did Dr. Pickering when you questioned him exclude Jupiter altogether?

Mr. MALLAN. I never questioned Dr. Pickering.

Mr. ANFUSO. You never did?

Mr. MALLAN. No.

Mr. ANFUSO. Where was Jupiter at the time?

Mr. MALLAN. Jupiter was from 13 to 16 degrees east of the moon. The signals, according to the chart from Goldstone—the ones that Dr. Dryden sent me—show the signals coming from the west of the moon. This is true. However, the director of Goldstone, Walter Larkin, told me that they were scanning and getting signals from both sides of the moon. There is also the possibility, according to Dr. Kraus, the radio astronomer at Ohio State, that ionized gas clouds in space could diffract advanced signals from Jupiter out of line, but perhaps not that far and perhaps as far. He did not really know because very little is known about these gas clouds. But this is a possibility according to him.

Mr. ANFUSO. Mr. Mallan, do you recognize the possibility that the receipt of signals by certain stations could not be made public owing to the existence of classified information, location of the station, type of equipment, and so forth? Do you recognize that?

Mr. MALLAN. I recognize that.

Mr. ANFUSO. Have you taken that into consideration in giving your testimony?

Mr. MALLAN. Yes; I have. In the case of Goldstone they were not operating under a standard 960 megacycles. They were not using their solid state microwave amplifier because they could not use it in the jerry-built system they had thrown together inside the big antenna in order to get down to 183.6 megacycles.

So I understand about the security involved, but I also know that no security is involved in taking the dish down to 183.6. In fact, they told me how they did it.

Mr. ANFUSO. Do you agree, Mr. Mallan, in theory at least, that pieces of evidence, none of which is conclusive in itself, could be conclusive in their cumulative effect—what we lawyers understand as the circumstantial evidence?

Mr. MALLAN. This is true if the evidence is cumulative. For the bursts of signals, they are hardly cumulative.

Mr. ANFUSO. You think the only cumulative evidence we have are these four bursts of signals?

Mr. MALLAN. No. I have heard of one other bit of evidence that is not precisely evidence either.

Mr. ANFUSO. What was that?

Mr. MALLAN. That was a tape made in Hawaii, a 5-minute tape—and I think it was even played back at the last hearing—of signals that were recorded by a tracking station in Hawaii, which was not identified.

However, I later found out these signals were received on 19.995 megacycles, way down below the Goldstone signals.

Goldstone was the only station in the free world that heard anything on 183.6 megacycles.

Mr. ANFUSO. And the station in Hawaii could not have heard these signals?

Mr. MALLAN. Yes, it is quite possible. This brings in a reasonable doubt that a rocket may have been launched and got 3,000 miles or 4,000 miles out and quit, because these signals were coming in very loud and very clear. After 5 minutes they quit.

Mr. ANFUSO. Those are the signals from Hawaii?

Mr. MALLAN. From Hawaii.

Mr. ANFUSO. If the Soviets did not make the shot which they claim to have made, how do you explain the data which they claim was obtained by lunik? This was published in the report of the Soviet Academy of Science, volume 125, No. 2, in 1959.

Mr. MALLAN. Has anyone translated this data? I have yet to see any data from the Soviet Union, because data means something more than merely generalized information.

Mr. ANFUSO. It is your statement, then, that this information has not been translated?

Mr. MALLAN. I have never seen a translation of it.

Mr. ANFUSO. Do you have the data?

Mr. MALLAN. I do not have the data, but I have another——

Mr. ANFUSO. I understand we have that information and we are going to have it translated.

Mr. MALLAN. That is fine.

Mr. ANFUSO. Does that meet with your approval?

Mr. MALLAN. It certainly does.

Mr. ANFUSO. Mr. Mallan, you may take your seat.

Mr. MALLAN. Thank you.

Mr. ANFUSO. Mr. Mallan, I must ask you these questions. I will delay these questions until we first hear from other members.

Mr. Karth, do you have any questions on what has already been brought out?

Mr. KARTH. Mr. Chairman, I am a little disturbed, not necessarily about the ideas that Mr. Mallan might have about lunik, but about his overall interpretation of Russian technology when he says that in all aspects of technology the Russians are so far behind us. I refer to the text that was read today by the gentleman preceding him when he says "miles and years," and yet this does not conform with the information that we have, not only from the opinion but from the visual inspection of some of our scientists who took a look at Russian equipment.

I would like to refer to the hearings we had some time ago, May 12, in fact, in which some of our scientists had the opportunity to view some of the sputniks. Not only did they view a sputnik as the Russians said they had sent it up, but they saw the pieces laid out so they could take a very good look at the instrumentation and the mechanism, at the works.

The scientists that we talked to claimed that this is equipment no one has to be ashamed of, that this equipment was very good equipment.

Did you see any of the equipment, for example, that the scientists referred to that they saw which was in the first three sputniks?

Mr. MALLAN. I saw the equipment that the Soviets claim were instruments in those sputniks. I would like to know how these scientists saw pieces. Were these pieces recovered by the Soviets from orbit?

Mr. KARTH. I am not going to be so technical as to suggest that they were recovered, or even so foolish as to suggest that, but I am suggesting that the scientists apparently saw duplications of the equipment that had gone into the making of these sputniks. Being scientists, I assume they weren't subjected to a big Red hoax.

I certainly have a bit more confidence in them than that.

As far as accuracy is concerned, I was attempting to refresh my mind on some of the data. It seems that Dr. Dryden and some of the other people who testified before our committee are in deep disagreement, to say the least, with you.

For example, they point out that the perigee altitudes of the three sputniks according to the final computations made in this country, were at 142 miles, 140 miles, and 135 miles, and that this took quite precise engineering and quite precise instrumentation to do this.

Compare these perigee altitudes attained by the three sputniks with those of our Explorers, for example. Explorer I was at 224, Explorer III at 121, and Explorer IV at 163, a difference of 142 miles. There was a difference of only 9 miles in the perigee altitudes of the three sputniks.

So that you should not be misled, it was explained that we have better instrumentation than this, certainly a more accurate guidance system. But because of the lack of power to boost up instrumentation which would be more accurate, it was impossible to get our best instrumentation into space.

What are your sentiments with respect to this 9-mile accuracy that they achieved in the perigee altitude of the three sputniks that they sent up?

Mr. MALLAN. I have exactly the opposite opinion of these scientists, even though I have respect for any scientists. I think the nine miles—actually it was 15, but that doesn't matter, that is pretty close, too—the closeness of the perigees of the three sputniks indicated to me that they had very poor guidance, that they merely shot things straight up and put them into orbit with autopilots, each one a little heavier, but maybe not as heavy as they claim. There is no way to check these weights. If their weights are accurate, then they have to have that much more propulsion. This is true.

But just assume, as these scientists undoubtedly did not assume, that the weights were very similar, in which case the same type of rocket with a little more fuel added to it or another fuel tank might have put three sputniks into orbit with perigees that close, requiring only simple autopilots controlled by radio from the ground.

Mr. ANFUSO. Will you yield there a minute, Mr. Karth?

Mr. KARTH. Surely.

Mr. ANFUSO. In other words, our scientists who arrive at these conclusions had to assume the weight of these different sputniks?

Mr. MALLAN. Yes, sir.

Mr. ANFUSO. And they had to take the word of the Russians for those weights?

Mr. MALLAN. Yes, sir.

Mr. KARTH. Isn't it true, though, that from the data we have so far as it relates to the size of the sputniks, a scientist who knows something about this particular equipment can fairly closely evaluate the weight?

Mr. MALLAN. Yes.

Mr. KARTH. Don't you give some credence to this possibility upon which these scientists have predicted a weight which, in effect, I think, pretty closely agrees with the weight that was announced by the Russians? Does this make sense to you at all?

Mr. MALLAN. It would make sense if we knew the tankage. No figures have ever been released by the Russians on the tankage, nor has anyone ever seen the rockets that launched the sputniks.

Mr. KARTH. If I recall correctly, the scientists in this country felt that they could very fairly closely evaluate the weights of the rockets that were used from things that they had seen in space.

Mr. MALLAN. Would you like my opinion on that?

Mr. KARTH. Yes; I surely would.

Mr. MALLAN. The things you see in space with optical equipment depend entirely upon the way light is reflected from those objects. This is the way optics works. You could have, say, a tennis ball and put long fins on it at right angles to each other of highly polished aluminum foil. If these fins were each 3 feet long and were at right angles, and there were two of them, one at the top of the tennis ball and one at the bottom, every time—or one on each hemisphere of the tennis ball—every time sunlight were caught or any light from a star even were caught from any direction by these fins, it would be reflected straight out because the light would, you know, just following the laws of simple optics——

Mr. KARTH. I think I understand what you are trying to say, sir.

Mr. MALLAN. Would be reflected straight out to an optical instrument on the ground which would make the object look 5, 10, 20, or 30 times the size it actually is. It is a tennis ball in this case.

Mr. KARTH. Do you have a faint suspicion that perhaps the Russians went to this length to confuse our scientists as to the size of the rocket they might have launched?

Mr. MALLAN. I have more than a suspicion, because one of our own tracking stations, a radar tracking station located on the Grand Bahama Island, which is part of the downrange tracking system from Cape Canaveral, the Air Force test missile center—for several days followed sputnik II in an attempt to get a configuration by scanning it with radar as it went by at a fairly low altitude. Each time it made a pass they were able to scan for 15 minutes. Adding up their results, they found that the configuration they were able to deduce from their scans by radar, scanning section by section of the satellite, did not conform to any known geometrical forms of our satellites or the Russians'. It did not conform to a sphere, to a cylinder, or to a cone. They conformed more closely to the fin-shaped corner reflector.

However, they seemed not to have deduced from this that the Russians might have gone to all that trouble just to fool us.

I just remembered another point. Prof. Alla Masevich, who is vice president of the Astronomical Council of the Academy of Sciences of the U.S.S.R. and who also is head of their optical tracking system or their program—if you can call it that—for the sputniks was one of the Soviet delegates to the eighth—I believe it was the eighth—International Congress of the International Astronautical Federation at Barcelona, Spain, during which time Sputnik I was launched. She told reporters there—who, of course, immediately descended upon her—that someone must have misplaced the decimal point, that the weight was probably 18.4 pounds and not 184 pounds. This was later corrected by the Kremlin, and I have since spoken with her myself and she now stands by the 184 pounds.

Mr. KARTH. One other question, Mr. Chairman.

Mr. ANFUSO. Yes.

Mr. KARTH. When Dr. Newell appeared before our committee here on May 12, the committee, of course, was interested in some of the allegations that you had made in your story which you propounded to be facts.

Dr. Newell referred to a book or an article, "Ionosphere and Positive Ions," and he had this to say. Let me just read it to you and see if you care to comment on this.

Mr. MALLAN. All right.

Mr. KARTH. He says:

> I would like to point out that the Soviet papers on the distribution of electron density in the ionosphere and on the number of positive ions were impressive. The Soviet rocket and satellite experiments have apparently provided the first quantitative data on the profile of the ionosphere above the maximum of the F–2 region.

Then he refers to page 3 again in the right-hand column. Mr. Townsend, who apparently was the author, points out that a mass spectrometer was flown in Sputnik III and a little later in that paragraph he quotes the results given from the first several orbits, which show that the principle positive ion is that of atomic oxygen mass 16, and the region from 250 to 950 kilometers.

He mentioned other things.

> I would like to point out these 950 kilometers was above anything that we had gotten at that time.

Since that time apparently we have. On the next page he points out the Soviets have flown a large number of their so-called meteorological rockets during the IGY from launching sites in the Antarctic, the Arctic, aboard ship, and in many latitudes.

The results on the temperature distribution of an altitude at 60 kilometers were present. In comment I would like to point out we are still working on the development of a meteorological rocket.

Then Mr. Teague asks a question:

> How do you know that statement is true?

Mr. Newell says:

> We have the papers. We have the results from their papers, and I brought along the packages to show you how many papers we got from this conference. In my estimation it is not possible for them to invent the data that they have here presented.

Then he compares these results with ours and concludes by saying:

The data show the same overall temperature distribution with altitudes as is shown in U.S. data. However, for the first time complete curves were presented, apparently from a large number of firings in a complete orbit, showing the seasonal variations in temperatures at each of the six 5-kilometer intervals between 25 and 50 kilometers.

This indicates to me, as a layman—certainly not as a scientist, because I am not one—that some of the things you have said, and especially your statement that in all aspects of technology the Russians are so far behind, are incorrect.

Would you care to comment on that?

Mr. MALLAN. Yes; I would. With all due respect to Dr. Newell—he is a very great scientist and a great expert in the ionosphere in general—I read the Russian papers he is talking about, as well as the American papers that he is talking about. Both appeared in the journal Jet Propulsion, the official journal of the American Rocket Society. These papers match each other very closely. The reason they match each other closely, as I have discovered the long hard way after having visited the Soviet Union, is that our papers are not classified in these fields—very few of them, at any rate, are. They are readily available to the Russians——

Mr. KARTH. Let me interrupt you right there. They had this information and had made it available to our scientists before we had the information as a result of our own data acquisition. The point I make is that our data after it was achieved then compared with that data that we had previously received from Russia.

Mr. MALLAN. I thought I was listening carefully, and I thought you had read from Dr. Newell that he said, "Their data, apart from the fact they have plotted these curves"—which can be done by any good mathematician from data, and we have never bothered to do this, which is beside the point.

I thought, though, Dr. Newell had said, "Their results compared with our results." I had the impression, in other words, that their results agreed with ours and not that ours came after theirs, because I am sure Dr. Newell, who has been in this field doing work ever since the first Viking rocket was launched by the Naval Research Lab, would know that we had meteorological rockets up there in the ionosphere for at least 10 years.

Mr. KARTH. Go ahead, Mr. Chairman. I want to take a look at this.

Mr. ANFUSO. Mr. Van Pelt?

Mr. VAN PELT. No questions.

Mr. ANFUSO. We do have another witness here who came all the way from Colorado, and we must hear him; and we must terminate for today at least at 5 minutes to 4. But we are honored this afternoon to have in addition to the special committee some members of my own standing committee. If they have any questions, I would certainly like to hear from them, Mr. Mitchell and Mr. Daddario.

Mr. MITCHELL. Mr. Chairman, I just wanted to sit in because of my interest.

Mr. ANFUSO. Mr. Daddario.

Mr. DADDARIO. I have just one question, Mr. Chairman. As I understand——

Mr. KARTH. Would the gentleman yield just one moment?

Mr. DADDARIO. Surely.

48438—59——11

Mr. KARTH. I had read this, that he had compared their results with ours after we had received our results. So apparently the way I had interpreted it was correct insofar as the testimony that was received from Dr. Newell is concerned.

Mr. MALLAN. Did he say "after" we had our results?

Mr. KARTH. He says then he compared these results with our results.

Mr. MALLAN. But this could mean we had the results before the Russians.

Mr. KARTH. Insofar as the meteorological rocket is concerned, he says we are still working on the development of that.

Mr. MALLAN. I mean particularly the data. Data are the key to this whole thing. The Russians never release data.

Mr. KARTH. All I can say is what he told our committee. He said we compared our results, our data, with the data that had been released by the Russians, apparently after we had accumulated our data.

Mr. MALLAN. I saw these same papers and read them very carefully in Jet Propulsion. I wish I had brought them along.

Mr. KARTH. I don't know if they are the same papers or not, but I do want to clear up the testimony of Dr. Newell.

Mr. MALLAN. All right; I am sorry.

Mr. DADDARIO. Mr. Mallan, when President Eisenhower made his statement concerning the lunik, you feel, then, he was basing that statement on either improper or erroneous information. Is that correct?

Mr. MALLAN. Improper information.

Mr. DADDARIO. When you went to Russia and you made this exhaustive study which you have talked about, with what, then, were you able to compare it from the standpoint of source information?

Mr. MALLAN. I spent 5 years or more on our own bases. I have written and published five books about our air research and development effort, our research into the upper atmosphere. I have lived at these places such as Holloman Missile Development Center; Sacramento Peak, where the Air Force does all of its solar and upper atmosphere—not all of its upper atmosphere research but all of its solar research, which is tied very closely into it.

I have talked with some of the key scientists in the various programs. Over these years I have seen their instruments and examined them carefully and had them explained to me by the men who use them daily.

This not only was true of optical and radio instruments; it was equally true of missile guidance systems, aircraft guidance systems, the way we carry through an entire research program to develop something new.

I had this background to compare. In fact, when I went to the Soviet Union I expected them to be ahead of us, and I was shocked, quite literally shocked.

Mr. DADDARIO. My point is this: You base your whole statement on your experience over this period of time you have mentioned. Where else within Russia and what other people of a completely objective nature do you have with whom you can compare your own information so that you can make a comparison as to sources? You only have your own point of view, your own observation. You have no knowledge, do you, whether or not there is information, whether or not there is equipment which you have not been allowed to see within Russia?

Mr. MALLAN. There undoubtedly is equipment I didn't see; but, on the other hand, when you see equipment at important research centers, and this equipment is so inferior to our own and so inefficient that it is shocking and sometimes laughable——

Mr. DADDARIO. Wouldn't you say, Mr. Mallan—if I could interrupt right there—that it is possible they could have accumulated equipment of this type so as to give you this impression?

Mr. MALLAN. Again, I must say that is very flattering that they would do this for me because it would take a terrific amount of energy on their part to move an entire radio installation which includes antennas and generators and all kinds of equipment that fill up a whole building—to move this out and move in some very poor equipment to make me think they are way behind us, while the Kremlin, on the other hand, is boasting they are way ahead of us.

Mr. DADDARIO. While we are talking about that from that standpoint, then, of flattering you, isn't it possible, without approaching it that way, that the scientists who have appeared before this committee and have reported to us that they have seen equipment which in their experience and with their background, too—and the background of many of these scientists is excellent—they have come to the conclusion that this is good electronic guidance equipment—isn't it possible that they might have seen equipment you were not allowed to see?

Mr. MALLAN. It is quite possible. However, I know that when you pin scientists down to actually analyzing what they saw in terms of what they know we have as a standard, if they were using our equipment as a standard they wouldn't take this attitude.

When they say it is good serviceable guidance equipment, the V–2 guidance equipment was good serviceable equipment, but it will never send an intercontinental ballistic missile to the United States.

Mr. DADDARIO. Mr. Mallan, you are trying to paraphrase what these people said or might have said, and it is not my conclusion from hearing them that that was their testimony. I would like to go into something else.

You talked about the fact that there could have been several sources which would have provided the type of sound which Goldstone received and which Dr. Pickering referred to in this committee. Isn't it part of their job from day to day as they conduct their experiments to eliminate this type of source from their findings? This is always above them, isn't it, from day to day?

Mr. MALLAN. No; it is not.

Mr. DADDARIO. These airplanes that you showed here fly over every day. The ham radio people are working every day. The police radios are working every day. These are sources which—certainly being experienced people in this area—they would have known about and would have been able to eliminate as a possible source.

Mr. MALLAN. I would like to explain they had never before listened on 183.6 megacycles. They threw a dipole antenna into the focal point of their big antenna, which is normally tuned to 960 megacycles. It also makes use of a solid state microwave amplifying system, which reduces noise level in ratio to amplification.

Mr. DADDARIO. Mr. Mallan, again you are giving conclusions which they might have arrived at. It is my recollection of the testimony on this particular point that this group had made exhaustive analyses of the sounds that they heard, and that they eliminated source after source as the possible source of the sound.

Some of them, as I recall, were these things which you list here as being the possible sources.

I get back finally to just one point. When the President of the United States made the statement and he had all of the services of the Government, all of these scientists, all of the intelligence agencies at his command, all of the places to check and doublecheck, good sources, bad sources--people to depend upon--isn't that type of information more reliable coming from such a broad area of information than that which you have coming really just from your own studies and not checked and doublechecked with this tremendous apparatus of Government information service at your beck and call?

Mr. MALLAN. I don't believe that what you are saying is necessarily true. I would like to quote some eminent scientists. I would like to quote first Dr. Gordon Little, who is an expert in ionospherics at Table Mesa, Colo. This is the Central Propagation Laboratory of the National Bureau of Standards. I would like to quote just two persons. I would like to quote Dr. Little first. This is a verbatim quote.

Mr. KARTH. What is he a doctor of?

Mr. MALLAN. I am sure that Dr. Shapley who is here could tell you quite a bit about him.

Mr. KARTH. You are quoting him. You must know something about him.

Mr. MALLAN. He is an expert in the ionosphere, in the physics of the upper atmosphere.

Mr. KARTH. Is this the field in which he has his doctor's degree?

Mr. MALLAN. I don't know.

Mr. KARTH. If you don't mind, the reason I would like to have an answer to that question is because, for all I know, he may be a doctor of medicine and then a self-styled expert in the upper ionosphere.

Mr. MALLAN. He works for the National Bureau of Standards at their most important laboratory as far as radio propagation is concerned.

Mr. KARTH. Could you give for the record what he has his doctorate in?

Mr. MALLAN. No; I cannot.

Mr. ANFUSO. We have a witness here, Dr. Shapley, from the Bureau of Standards, who I am sure will be able to tell us.

I must ask you these questions, Mr. Mallan——

Mr. MALLAN. I was going to quote Dr. Little. When I asked him about whether or not they at his station had tried to record signals from the lunik on the lower frequencies, whether they had received any signals—and they have a radiotelescope inteferometer system there--he said: "We certainly didn't. I have no doubt that other groups in the U. S. did, since the radio transmitting frequencies were very firmly announced."

To follow that I would like to quote Dr. Harold O. Curtis. I don't know what his doctorate is in, but I do know that he is associate director of Project Space Track at the Air Force Cambridge Research Center, the project that correlates all the military tracking data in the free world and analyzes it and decides where to send it for best use.

Mr. ANFUSO. Is this the only quotation you are going to make?

Mr. MALLAN. This next one is the last one. Dr. Curtis told me that two stations had reported signals to him. One was the Goldstone tracking station on one frequency, the higher frequency; the other was the Stanford Research Institute station at Menlo Park, Calif., on the lower frequency.

Then I asked him: "Were the signals loud and clear?" He said: "Not exactly." After the discussion, it turned out they were weak and sporadic and heard only for a brief period. Then I asked him: "Are the Jet Propulsion Laboratory, Goldstone, and Stanford certain with absolute scientific certainty that these signals were coming from a vehicle heading toward or past the moon?"

His answer was verbatim: "If you mean could the signals be coming from some other source, the answer is yes."

His analysis of these signals is now a classified document in the Pentagon. I don't see why signals that are definite should be classified. We never classified signals from the sputnik. Why from the alleged lunik?

Mr. ANFUSO. Mr. Mallan, have you traveled to Europe on any occasions prior to your trip to the Soviet Union in 1958?

Mr. MALLAN. Yes, I have.

Mr. ANFUSO. When was that?

Mr. MALLAN. That was in 1937, August.

Mr. ANFUSO. Did you obtain a passport in 1937?

Mr. MALLAN. Yes, I did.

Mr. ANFUSO. What countries did you represent that you desired to visit?

Mr. MALLAN. I visited Spain at the time.

Mr. ANFUSO. Is that the only country you stated you desired to visit?

Mr. MALLAN. No. I stated I desired to visit France. I didn't state that I desired to visit Spain.

Mr. ANFUSO. I see. The only statement you made on your passport was you intended to visit France?

Mr. MALLAN. That is right.

Mr. ANFUSO. But you wound up in Spain?

Mr. MALLAN. That's right.

Mr. ANFUSO. What was your address at the time you made this application?

Mr. MALLAN. It was on Forward Avenue in Pittsburgh, Pa., but I forget the exact number.

Mr. ANFUSO. Did you surrender the passport on your return to the United States?

Mr. MALLAN. I surrendered my temporary passport on return to the United States. My original passport was taken away from me in Spain.

Mr. ANFUSO. When was that taken away?

Mr. MALLAN. About the end of August 1937.

Mr. ANFUSO. What name did you use on the original passport?

Mr. MALLAN. My own name.

Mr. ANFUSO. Spelled the same way?

Mr. MALLAN. Spelled exactly the same way.

Mr. ANFUSO. You surrendered your passport—that is the original passport; is that right?

Mr. MALLAN. You mean the original one that was taken away from me in Spain?

Mr. ANFUSO. Yes.

Mr. MALLAN. I didn't want to surrender it, by the way.

Mr. ANFUSO. You were asked to surrender it?

Mr. MALLAN. They told me they would return it to me.

Mr. ANFUSO. Who is "they"?

Mr. MALLAN. Some people at Barcelona, Spain.

Mr. ANFUSO. Were they officials of the U.S. Government?

Mr. MALLAN. No, they were not.

Mr. ANFUSO. Was it stolen from you?

Mr. MALLAN. I later learned in a sense that is what it amounted to.

Mr. ANFUSO. How did you get back to the United States?

Mr. MALLAN. The U.S. consulate in Barcelona gave me a refugee passport, a temporary passport, and drove me into France.

Mr. ANFUSO. What year was that?

Mr. MALLAN. That was in 1938 in, I believe it was, April, early April.

Mr. ANFUSO. What was the exact date of your departure from this country and the exact date of your return to this country?

Mr. MALLAN. The exact dates at the moment I can't remember.

Mr. ANFUSO. Could you give us the months?

Mr. MALLAN. It was about the middle of August that I left the country, 1937, and returned about the middle, I think, or the end of April 1938.

Mr. ANFUSO. Are Spain and France the only two countries that you visited while you were abroad?

Mr. MALLAN. Those are the only two.

Mr. ANFUSO. Did you indicate on your passport your intention to visit Spain? I think you said a moment ago you did not.

Mr. MALLAN. No, I did not.

Mr. ANFUSO. Did you know before leaving the United States that the real purpose of your travel was to eventually land in Spain?

Mr. MALLAN. Yes, I did.

Mr. ANFUSO. What was the purpose of your going to Spain?

Mr. MALLAN. The purpose was an idealistic one.

Mr. ANFUSO. Will you state it?

Mr. MALLAN. Yes. I thought a war that was being fought there at the time was a war against fascism. I didn't really feel like being shot at, but I had been a radio operator in the Naval Reserve and I knew quite a bit about radio. There was an organization called the North American Committee for Technical Aid to Spanish Democracy which I applied to as a radio technician.

I wound up as a machinegunner against my will in a Canadian battalion, the McKenzie Papinal Battalion.

Mr. ANFUSO. What was that?

Mr. MALLAN. McKenzie Papinal Battalion.

Mr. ANFUSO. That was a Canadian battalion?

Mr. MALLAN. That is right. However, I later managed to get transferred to Valencia as a radio technician. By this time I was thoroughly disillusioned with the idealistic war and so I found my way to Barcelona, which is a seaport. I tried to stow away on a British freighter, the R.M.S. *Stanhope*, and some British seamen found me in the boilerroom and turned me over to the carabinieris, who turned me over to the assault guards who—with no charges against me by the way—had me listed for expulsion from the country. Meanwhile, I got 3 months in Carcer Modello, the model jail of Barcelona.

Mr. ANFUSO. How old were you at the time?

Mr. MALLAN. I was 22 when I left and—22 and a half or something like that.

Mr. ANFUSO. How old are you now?

Mr. MALLAN. I am now 44, en route to 45.

Mr. ANFUSO. Did you serve in the International Brigade?

Mr. MALLAN. Yesterday afternoon I discovered that I had served in the International Brigade because the McKenzie Papinal Battalion a few weeks, I believe it was, after I had been placed in it, had been consolidated into the Abraham Lincoln Brigade, that is. Originally I had not thought I was in that brigade until yesterday afternoon.

Mr. ANFUSO. You say you were absorbed into the International Brigade. Is that it?

Mr. MALLAN. I wasn't. The battalion——

Mr. ANFUSO. The battalion was absorbed?

Mr. MALLAN. That I had been placed in; yes.

Mr. ANFUSO. That was the Canadian battalion?

Mr. MALLAN. Yes.

Mr. ANFUSO. It was absorbed in the International Brigade?

Mr. MALLAN. No; it was a part of the International Brigade because it was absorbed into the Lincoln Brigade. But, meanwhile, my original intention had been to go as a radio technician and be no member of any battalion or any brigade.

Mr. ANFUSO. Do you know anything about the Abraham Lincoln Brigade?

Mr. MALLAN. No; not really.

Mr. ANFUSO. Do you know it would be on the Attorney General's Communist list?

Mr. MALLAN. I learned this yesterday also.

Mr. ANFUSO. Only yesterday?

Mr. MALLAN. Yes.

Mr. ANFUSO. You never knew that before?

Mr. MALLAN. No; I did not.

Mr. ANFUSO. You were born in this country?

Mr. MALLAN. I was born in Pittsburgh, Pa.

Mr. ANFUSO. Where was your father born?

Mr. MALLAN. I think he was born in Poland. I am not sure, which is now a part of Russia, that part of Poland.

Mr. ANFUSO. But at that time it was not part of Russia, though it is part of Russia now—is that correct?

Mr. MALLAN. That is right.

Mr. ANFUSO. Can you tell this committee why you concealed from the State Department the true purpose of your trip?

Mr. MALLAN. All passports were stamped restricted for travel in Spain. I realized that I broke a law when I went into Spain.

Mr. ANFUSO. Did you feel, Mr. Mallan, that if you had stated that you wanted to go to Spain for the purpose of fighting with the Loyalists against the Fascists, that you would never have been given a passport?

Mr. MALLAN. I didn't want to go to fight. I wanted to go as a radio technician.

Mr. ANFUSO. Even if you had stated that purpose, you feel that you would never have been given a passport, is that it?

Mr. MALLAN. That is right.

Mr. ANFUSO. Since then you say you regretted that move?

Mr. MALLAN. I have certainly and deeply regretted it.

Mr. ANFUSO. By the way, for this particular service were you recruited by anybody?

Mr. MALLAN. Yes; I was recruited.

Mr. ANFUSO. By whom?

Mr. MALLAN. I don't recall an individual's name. There were many individuals.

Mr. ANFUSO. Were you recruited in this country?

Mr. MALLAN. Yes.

Mr. ANFUSO. And you don't recall by whom? Was it an individual?

Mr. MALLAN. It was probably several individuals.

Mr. ANFUSO. Had you ever seen these individuals before?

Mr. MALLAN. Yes, I had.

Mr. ANFUSO. Before you were recruited?

Mr. MALLAN. Yes.

Mr. ANFUSO. How many times had you seen them before?

Mr. MALLAN. I would say more than a few times.

Mr. ANFUSO. How many people would you say that you had met before who were finally responsible for your being recruited in the service?

Mr. MALLAN. Half a dozen.

Mr. ANFUSO. Do you remember any of their names?

Mr. MALLAN. I can remember one of their names. I would prefer to give it to you in private if that is permissible.

Mr. ANFUSO. Only one that you can remember?

Mr. MALLAN. In this respect, in the respect of recruiting me. But there were more than one.

Mr. ANFUSO. But you can remember only one name?

Mr. MALLAN. No; now I remember another one too, which I will also give you in private.

Mr. ANFUSO. You have no objections to putting those names down for the record?

Mr. MALLAN. I have no objection.

Mr. ANFUSO. Did any of these individuals tell you how to get a passport, or did they furnish the passport for you?

Mr. MALLAN. No; I applied for the passport myself through regular channels.

Mr. ANFUSO. Did they tell you what to say in order to obtain this passport?

Mr. MALLAN. No, they did not.

Mr. ANFUSO. Were they Loyalists? Were they sympathizers too?

Mr. MALLAN. I don't think they—you mean Spanish Loyalists?

Mr. ANFUSO. Spanish Loyalists.
Mr. MALLAN. No, they were not.
Mr. ANFUO. That you are positive of?
Mr. MALLAN. Yes.
Mr. ANFUSO. To the best of your recollection they were not Communists?
Mr. MALLAN. One of them was a Communist.
Mr. ANFUSO. One of them was a Communist?
Mr. MALLAN. Yes.
Mr. ANFUSO. Is he still a Communist?
Mr. MALLAN. He is dead.
Mr. ANFUSO. Who paid for your transportation?
Mr. MALLAN. The North American Committee for Technical Aid to Spanish Democracy, as they phrased it.
Mr. ANFUSO. Mr. King.
Mr. KING. Mr. Mallan, we have had before us Dr. Pickering, associated with the Goldstone tracking station some 2 weeks ago. He had all of his entourage with him, 6 or 8 or 10 men as far as I could tell, from Goldstone. When he was here I remembered your statement and your original article in which you stated in substance, I believe, that you had made inquiry of all those Government officials that you felt would be concerned with tracking this lunik, if there was a lunik. At least you had made checks with the keymen here and there. I don't want to misquote you, but I am just remembering the substance——
Mr. MALLAN. No; with many keymen.
Mr. KING. I got the impression that you stated that you contacted what you considered to be all the keymen who would be in a position to have tracked lunik if there had been a lunik.
Mr. MALLAN. I didn't state that.
Mr. KING. Am I overstating it? All right, I am just really relying on my memory. In any event, I did have occasion to ask Dr. Pickering whether he or any member associated with him at Goldstone had been contacted by you. Their answer was "No," not until after the article appeared; that you had made contact with them, but not prior to the publication of the article.
My quesion is, is that a true statement as far as you are concerned, and if it is, why did you not contact them before you wrote the article?
Mr. DAIGH. Might I interrupt a minute because I have the exact account you referred to and it might be more accurate to read it.
Mr. KING. I wish you would.
Mr. DAIGH. Mr. Mallan is speaking of the occasion after lunik was allegedly fired. He says, "I couldn't just let the incredible thought ferment in my brain. I went to Washington, talked with military men and intelligence officers in the Pentagon. I visited projects, Space Track, the Air Force installation in Massachusetts, that correlates tracking data from all over the free world. I telephoned major tracking stations. I talked with scientists."
Mr. KING. That is probably what I had in mind when I said you telephoned major tracking stations. I interpreted that to mean all the major tracking stations. Maybe I am reading more than I was justified, but go ahead and answer my question.

Mr. MALLAN. I believe the day before publication day, or several days before, I spoke with Walter Larkin for the first time. He is the director of Goldstone. I am not quite sure of this until I check back on my own dates, but it appears to me that I did.

During that hour-long conversation it boiled down to the fact that he was fairly sure that every time I asked him could the signals have been coming from, say, television reflections or some other source, quoting him almost exactly, he said:

They could have been coming from television or almost anything, but Dr. Pickering reduced the data and Dr. Pickering is sure that the data is correct. If it is all right with Dr. Pickering, it is all right with me because he is the man who pays my salary.

Mr. KING. This is what he told you over the phone?

Mr. MALLAN. Yes.

Mr. KARTH. Will the gentleman yield?

Mr. KING. Yes.

Mr. KARTH. According to the transcript, Dr. Pickering says that you talked to no one in the laboratory prior to the publication.

Mr. MALLAN. Not at JPL, but at Goldstone itself, which is near Camp Irwin in the Mojave Desert. I talked directly to the Goldstone station, to Walter Larkin there, the director.

Mr. KING. Mr. Mallan from what you have just said, do I not detect an apparent inconsistency? Maybe not. If not, I wish you would correct me. But your article certainly suggests to the reasonable mind that you canvassed the customary sources, checking most, if not all, of the major stations, and that nothing you found was in anyway inconsistent with the conclusions you had reached. Yet your statement just a moment ago is that Dr. Pickering's computations seemed to indicate that there was a lunik, and if that is the way it was, that is the way it was.

Wasn't that enough notice to you, just on the basis of what you said a minute ago, to suggest that there was something quite inconsistent with the conclusions you had reached?

Mr. MALLAN. No; because I haven't told you the entire telephone conversation.

Mr. KING. Would you like to do so?

Mr. MALLAN. I believe I have a transcript of at least a good part of it here.

Mr. KARTH. As long as you have the transcript, you might also give us the exact date of this conversation.

Mr. MALLAN. Yes; I have—I hope I have that transcript.

Mr. KING. Do you find it in the material you are looking for, Mr. Mallan?

Mr. MALLAN. Pardon me?

Mr. ANFUSO. Do you find the material you are looking for?

Mr. MALLAN. Yes; I am also trying to find the date for you, though, the date which I first talked with Walter Larkin.

Mr. KING. Is that a lengthy statement?

Mr. MALLAN. Well, it is just quotations. It would take 2 minutes, perhaps. I would like to point out, first, however, that I did not know Dr. Pickering's attitude about the whole matter at any point along the way until Dr. Dryden made it public, after the article did appear, or I would certainly have tried to get in touch with Dr. Pickering.

I did try to get in touch with Dr. Pickering afterward—I mean after the article appeared. His secretary recognized my name and said she would ask him to phone me back.

He was expected in another hour or two, she said.

I purposely waited at the office until long after 5 o'clock because of the time differential, and he did not phone me back then, nor did he phone me back the following day, nor has he ever phoned me back. That is beside the point, I guess.

Mr. KING. Would you like to read the material?

Mr. MALLAN. Yes; I asked Walter Larkin, the director of Goldstone, how strong the signals had been:

> They varied from medium weak to nonexistent.

That was his answer. Quote again:

> We were never swamped with them.

Next I asked him for how long a period the signals had been heard.

> The longest time was probably in the neighborhood of 3 minutes—

he said—

> and then they would fade and reappear 4 or 5 minutes later.

How long was the total time the signals had been tracked?

> We tracked sporadically from about 2 a.m. to about 7 or 8 a.m.—

he answered.

Could you say with absolute certainty that the signals were coming from a Russian vehicle moving beyond the moon? I asked him.

> I would have to say that Dr. Pickering made a statement to the press that Goldstone tracked a Russian missile.

But you are sure the signal was intelligently modulated?

> Dr. Pickering said it was. He pays my salary. If he is sure, it is good enough for me.

You say the lunik signals were extremely weak, yet the Russians claimed they were coming in clearly. I have seen their equipment and it is not nearly as good as yours. How do you explain this?

> They may have some better equipment hidden away.

If they do, it is so well hidden that their own scientists cannot find it and use it. Let me tell you a significant incident.

I told Mr. Larkin about an anecdote published in the January 1958 issue of Space Flight, official publication of the British Interplanetary Society. A member of the research team connected with the great radiotelescope at Jodrell Bank was describing a situation he had experienced with Sputnik No. 1. This is the research fellow at Jodrell Bank:

> Three weeks after the satellite was launched—

he wrote—

> the batteries feeding the satellite's own transmitter went to sleep and prevented further radio observations throughout the world. Apart from visual sightings, the satellite was lost unless radar could again detect it. This would have included the Russians themselves. A telegram arrived at Jodrell Bank posthaste from the Soviet Academy of Sciences asking if the telescope could

be used to fix on the rocket. Within a few hours the fix went back giving the exact range and time of transit over the Mediterranean—figure which confirmed exactly the British computations of the orbit, but which were significantly different from the Russians' own predictions.

The next question I asked of Mr. Larkin was designed to find out whether or not there were any radioastronomers among the Goldstone tracking station personnel. Normally I asked him "Does your station do any research work in the field of radioastronomy?"

We are basically a tracking research station. We do not do radioastronomy work.

Next I asked him "I think this whole lunik affair was a magnificent hoax. I am absolutely certain of it."

Walter Larkin answered this as follows:

I think this sentiment is not only yours. You are not the only one who feels that way.

I asked him quickly after that:

Who else do you know who has my attitude?
I believe it was Drew Pearson who came out with a column on it.
No, I meant have you heard any scientists say this?

There was a long pause and then Mr. Larkin answered "No comment."

That was the first conversation I had with him.

Mr. KING. Do you find the date in your own notes of that conversation?

Mr. MALLAN. No, it was before April 15 and I will find the exact date. I apparently left it at home. This is the retyped copy.

Mr. KING. What is the significance of the April 15 date?

Mr. MALLAN. The magazine did not appear until April 21.

Mr. KING. You said it was before April 15. How do you fix that in your mind?

Mr. MALLAN. Because of the big telephone bill he got which had to be before April 15.

Mr. DAIGH. There was a $300 phone bill.

Mr. KARTH. When was the first date you made an advance release, not for general publication, but the first date the story was released?

Mr. DAIGH. It was on the Friday before the 21st, is my best recollection, which I think can be confirmed.

Mr. KING. When was the article sent to the printer?

Mr. DAIGH. I would have to answer that: I think we have some wrong dates. This went on sale in March, did it not? The April issue or the May issue?

Mr. MALLAN. That is the May issue.

Mr. DAIGH. It went on sale in April.

Mr. MALLAN. April 21. I do not know when it was sent to the printer myself.

Mr. KING. You had written your article, though, when this conversation was had?

Mr. MALLAN. Pardon me?

Mr. KING. Your article had already been written?

Mr. MALLAN. Oh, yes, the article had been written. I was trying to verify what I had found out about the Goldstone signals at Spacetrack.

Mr. KING. In fact, you had submitted your manuscript, had you not?

Mr. MALLAN. We do not work that way.

Mr. KARTH. Your conclusions had been drawn. Let us put it that way.

Mr. MALLAN. My conclusions had been drawn.

Mr. DAIGH. But he was verifying it, if I may add. He was verifying the information that had been given him in Massachusetts. As is common, he was checking another source. Goldstone was the source mentioned in Massachusetts, so he called there.

Mr. KARTH. As a result of any information that you got from any Government agency, tracking or otherwise, did you change any of your original text?

Mr. MALLAN. My original text was based upon information that I had gotten from tracking and Government correlating sources.

Mr. KARTH. Those that you did check appear in the article?

Mr. MALLAN. They are not named because in the article I promised not to name them.

Mr. KING. Do you have any more questions?

Mr. KARTH. I have one more question I would like to ask. I do not agree with you on this.

Mr. Daigh, you say on page 4 of your testimony:

> It is inconceivable that scientists would lie about their lifework and specialties, whether it be astronomy, space tracing, and missile guidance.

I do not quite agree with that statement. I think instead of its being inconceivable it is entirely conceivable that scientists or members of any other profession might lie, if the Russian Government said you must not divulge this information or you must report erroneously on something else. I think it is entirely conceivable rather than inconceivable that they might do so.

What is entirely inconceivable in my opinion is that the American officials, military and nonmilitary, who have appeared before this committee and who have stated their opinions on this subject matter are lying to us about what they seem to know are the facts and what they seem to speculate. So I would certainly have to disagree with you on this particular point.

If you want to comment further, it is fine with me. But I disagree that it is inconceivable they would lie.

Mr. DAIGH. You are referring to my testimony, I believe.

Mr. KARTH. Yes.

Mr. DAIGH. I would like to point out first the hierarchy of science in Russia is nonpolitical, and there are facts that would show that most of the major scientists are not even members of the Communist Party.

I can speak only from my experience as an editor and a reporter. It is by belief, and you are certainly privileged to differ with me, that a trained observer interpreting 38 scientists, talking to them about their specialties, asking them about their theories, examining their equipment, is going to—in conversations that ran 3, 3½ hours—detect a falsehood if there is one. Let us say he misses one; but if there is a general attitude of falsehood, that is what a reporter is trained to detect.

In my own experience I have written considerable science. I would go flatly with this statement that any scientist I have ever talked to, in my belief, would not willingly lie about his specialty. His whole reputation is in his specialty.

Mr. KARTH. Let me ask you this: Do you think the American scientists who appear before this committee, with and without the jurisdiction of the American Government, are lying just because they disagree with the conclusions that were drawn by Mr. Mallan?

Mr. DAIGH. Definitely. The human race is at best only reasonably efficient.

Mr. KARTH. Yes, sir. But you see, the committee is in a position where they either have to agree with Mr. Mallan or they have to agree with all of the scientists who appeared before this committee. If you want to carry it to the ultimate and say they are lying—I do not want to say that; it might be a difference of opinion. Then, if we are going to conclude by saying one of the two groups is lying, we would have to, on the basis of the testimony received by this committee, say that Mr. Mallan is lying.

Mr. DAIGH. Let me say this: I am very happy that it is not a committee of scientists who are going to resolve and make this decision. I like to play poker. I love to play poker with scientists. I would hate to play poker with Congressmen. I am sure that these men are not lying. They are giving their best answer, absolutely their best answer. But I think that you gentlemen, on the evidence given, are far better able to make a determination of fact than a little group of men who have clustered themselves around Goldstone, Calif., and say "This one station in the whole world is sufficient to prove the existence of this rocket." I do not know whether you were here when we showed that dozens of ham operators in Russia were reported in the Russian press as having received this signal from this rocket for durations of hours, up 280,000 miles.

Not our finest equipment could locate it.

Mr. KARTH. Of course, when you are talking about our finest equipment now, it seems to me I recall the testimony given to this committee that when they received information about the firing of this lunik they did not—if I recall correctly—have sufficient time to arrange for their best possible equipment to be in operation at the time of the firing so that they could more accurately and more adequately perhaps track——

Mr. DAIGH. At the time of the firing, that is correct.

Mr. KARTH. Because of the insufficient time that was given and everything was helter-skelter, and they did do this with inferior equipment, in effect.

Mr. DAIGH. I know there has been testimony to that effect. I do not understand why it should have to be in a closed hearing. We are here to determine the truth. If you have proof after you walk out of that door, I think the least you can say is that we have put a very reasonable doubt on this, no matter who testifies in what direction.

I would like to say this, that Jodrell Bank was satisfied. That is the finest piece of that type of equipment in the world. They sent a telegram, a cable to this committee, in which, if I remember it correctly, they said they were precisely tuned and were baffled, could not understand it. We have had similar messages from Japan and France. There is no logical explanation for this.

I am sorry that I have to place myself against such expert testimony as the men who have told you to the contrary. But I still can only evaluate what I can see and make a measurement and vote the way I believe is right. That is your privilege.

Mr. KARTH. Certainly. I appreciate that. I am not attempting to harangue about the fact that you are here testifying. The fact is, I appreciate it as an individual and as a Member of Congress.

The only point I make is that I do not agree with you that it is inconceivable that the Russian scientists would lie either by design, by desire, or by order; and by so stating, you imply quite oppositely that those people who are conscientious American governmental workers, if you will, officials to be sure, and scientists, in effect, then, perhaps, are lying on the basis of what they have seen of Russia.

These people have been there also, and I am sure that you are familiar with that.

Mr. DAIGH. Yes.

Mr. KARTH. Many American scientists have been to Russia. Many American scientists have talked to people who Mr. Mallan has not talked to. The fact is, there are some American scientists who say that Mr. Mallan did not talk to top Russian scientists. In and among the 38 he talked with, of course, there are some who are quite creditable in their scientific professions, but he did not talk with the real top scientists.

Again, those American scientists who appeared here, who have been to Russia, have stated, apparently without fear of contradiction, that they as scientists have viewed, visually seen, pieces of electronic equipment and guidance control equipment that is on a par in many instances with ours, and this does not indicate to me that the Russians are years and miles behind us in the field of technology.

This is the point I am attempting to make, Mr. Daigh.

Mr. DAIGH. Yes, I understand that.

Mr. DADDARIO. Will you yield, Mr. Karth?

Mr. KARTH. I am finished.

Mr. DADDARIO. Mr. Daigh, to get after this just a bit more, this paragraph which Mr. Karth has called your attention to—your answer to it, as I recall the first part of it, was that these people are not members of the Communist Party and therefore you lay great credence on their being extremely truthful.

How much do you know of the background of all of the people to whom Mr. Mallan spoke? How do you know they have been affected by the Russian state and the police system which is certainly in existence there; and how do you in fact know whether or not they are members of the Communist Party or of some secret branch thereof? How could you possibly have the information to have complete backgrounds of these people so that you could make such an objective analysis and come up with this conclusion that it is inconceivable that these people would lie?

Mr. DAIGH. I do know that only a small statistical percentage of the entire Russian country are members of the Communist Party. I do know from my reading that there is a notable absence among the higher scientists as members of the Communist Party. I cannot state specifically that the men Mr. Mallan saw were or were not. I would believe, though, that it would be a mathematical certainty that some of them were not.

However, more to the point, 38 men, scientists, are not members of the Russian Ballet or the Russian Art Theater, and from my experience would make the lousiest actors in the world. To give them all the same party line that this one writer, this little writer of books—because that is all Mr. Mallan was known to them as; they were told he was going to write magazine articles; he was over there to write a book—that for this one little man, an extravagant show of thespian ability from an unsuited audience or an unsuited participant is an absolute impossibility in my opinion.

Mr. DADDARIO. Well, Mr. Daigh, we would follow that through logically then, and if we are to believe every single thing which Mr. Mallan has said, then we don't have really any worry or we are in no rush insofar as our science program is concerned to keep ahead of the Russians. We are already way ahead of them. We already have progressed far enough and therefore we need not spend all this money and we might as well stop now.

Is that what you mean?

Mr. DAIGH. I certainly wouldn't agree with you.

Mr. DADDARIO. I am not making the statement that I think we should. I am saying that if we follow your line of reasoning we certainly are way ahead in this race with the Russians and we need not be too much concerned about keeping pace with them or catching up.

Mr. DAIGH. To my own opinion I think that attitude would be the height of stupidity, because I am equally convinced that if the Russians had the means they would drop a bomb on Washington tomorrow.

Mr. DADDARIO. If that is the case—if I might interrupt right there—if that is the case and if it would be stupid and if an article of this type might tend to slow down the whole operation of the space missile program in this country, wouldn't it be conceivable under those circumstances that the Russians could get 20 or 30 people and have them tell a story to this effect? Because certainly it could be a disastrous kind of thing if we were to believe this, and if we were to match our program up against theirs on the basis of this statement and then necessarily, because there would not be any need of keeping ahead with such speed and spending such money, slow down our whole program?

Mr. DAIGH. I would believe that if any such propaganda effort were made it would be quickly recognized because it would be so in contradiction to the well-known party line of propaganda that has emanated from Moscow since shortly after the revolution in which they have claimed with sober face to have invented every accomplishment that the free world actually invented, from the airplane to the electric light to the internal combustion engine, penicillin, television, radio. They haven't missed one. They haven't missed a single effort to boast.

Now to conceive that they would suddenly switch and use this man as the only instrument or the wedge is inconceivable. However, I will say again if this happens to be the case, it would be, I think, easily detectable.

Mr. KARTH. Would the gentleman yield?

Mr. DADDARIO. Yes, sir.

Mr. KARTH. Perhaps they were not trying to get Mr. Mallan to write his article in the vein in which he did write it, but certainly they may have been instructed to give Mr. Mallan very little, if any, accredited information on the basis of orders that they had received. Maybe from that information that they did give Mr. Mallan, he erroneously drew his conclusions. Would you say there is a possibility that that happened?

Mr. DAIGH. I gave that a great deal of consideration before this article was published—a great deal of consideration. I have the same conclusion I had at that time, that it would have been physically and literally impossible. You must remember that this man walked into laboratory after laboratory and into observatory after observatory.

He made many of his conclusions not on what they told him, but what he saw. He has a comparison base to start from. He has been in similar equipments in this country. For example, when he saw a quartz clock and was told that this was Russia's finest way of splitting seconds, there was no reason really to doubt them. They had it hooked up to their optical tracking system.

It was interesting and typical of their heavy-handed engineering that they had attached a printer to it which cut its efficiency over 60 percent. While we can use a quartz clock for accuracy within a thousandth part of a second, they were boasting that their accuracy was around three hundredths of a second.

This equipment, incidentally, was manufactured in the United States of America.

Mr. KARTH. Certainly they are not suffering under the delusion, I suppose, that our quartz clocks are not sufficiently accurate so as to be able to determine accuracy within a thousandth of a second; are they? Certainly they understand this; don't they?

Mr. DAIGH. Mr. Mallan could answer that better, but I will try to do it. Several of the scientists told him——

Mr. KARTH. Not that I think it is of any importance, because I think we are ahead of them in automobiles, too, if you want to talk about nice, shiny, chromy automobiles and a number of them on the streets. There is no argument. We are way ahead of them. We are way ahead of Russia in many things.

Mr. DAIGH. These scientists told Mr. Mallan, for what it is worth, that they understood we were working on atomic clocks, doing theoretical work. There was no observation that he could make in his talk with numerous of these scientists that they knew we had it at the time. Today it is being mass produced.

Mr. KARTH. This is kind of general knowledge about the ability of our quartz clock.

Mr. DAIGH. We have got a finer machine called the atomic clock. We wouldn't depend on the quartz clock for missile firing today. The atomic clock can split a second into a millionth.

Mr. KARTH. Certainly the Russians know about that, don't they?

Mr. DAIGH. They must now because it is being mass produced up in New England. Mr. Mallan can tell you that better. Would you do it?

Mr. MALLAN. Sure.

Mr. KARTH. I think you are doing a very good job.

Mr. DAIGH. I wasn't there.

48438—59——12

Mr. MALLAN. You mean tell him about the atomic clock?

Mr. DAIGH. Tell him what you found out as concerns the Russian use of precise timing.

Mr. MALLAN. Oh, yes. Precise timing is very necessary, especially to launch a lunik.

Mr. KARTH. Let me ask you this. Were you in a station that they would probably be using as the best station for tracking the lunik or the sputniks or anything else?

Mr. MALLAN. It was supposedly the headquarters station for tracking the sputniks. It was the Sternberg Astronomical Institute, which is the astronomical institute of the Academy of Sciences, U.S.S.R. Alla Masevich, who is quite well known in this country and who is head of the sputnik optical tracking system, took me upstairs to the roof of the institute and proudly showed me the setup after having showed me the time--what did she call it—the time—we call them time signal generator rooms, but she called it the time standard room, that's right. It was the most haywire conglomeration of electronic equipment I had seen in a long time.

Mr. KARTH. Do you think this represented their finest electronic equipment? Is that correct?

Mr. MALLAN. I didn't say that.

Mr. KARTH. I am asking you a question.

Mr. MALLAN. I said this was the important astronomical institute, and is was the time standard room.

Mr. KARTH. Do you think this represented their finest technological advances in this field?

Mr. MALLAN. I can't answer that.

Mr. KARTH. Do you think so?

Mr. MALLAN. No, they must have better things than that. But it still was the time standard room and they were using American crystal clocks which they were able to purchase from us because of the IGY.

They had devised and were very proud of this mechanical printer which translated electronic impulses into mechanical impressions on a paper strip, thereby losing a lot of efficiency, plus the loss in efficiency caused by fluctuations in their powerlines and no air conditioning, and you have a situation where a clock that just normally operates at an accuracy of one one-thousandth of a second operating at from one two-hundredths to one three-hundredths of a second, and Alla Masevich was very proud of this.

Mr. KARTH. Let me ask you this question——

Mr. MALLAN. I think that is significant, don't you?

Mr. KARTH. Let me say this: I think it is significant to the point that at least in all stations the Russians probably are not on a par with us in the installation of the very most modern equipment. If they had a great deal of new modern equipment, then I would assume that they would also have had it there.

Mr. MALLAN. Yes.

Mr. KARTH. But it doesn't necessarily convince me that because they didn't have it there that they don't have it. That is the point I make. To make my position perfectly clear, I would very much like to have your story be true, and I am sure that every Member of this Congress would like to have this story of yours be true. I am sure that all of the American people and the people in the freedom-loving countries all around the world would like to have your story be true.

So our attempts to get information from you are not meant to be of a derogatory nature, and what we say doesn't necessarily profess our "druthers," if I may use the word. We would rather, as Members of Congress and as Americans, be perfectly confident that your story is true and that we are years and miles ahead of Russia in all fields of technology. This would be wonderful. This would be the most tremendous news that our people have had for years, at least, wouldn't it?

But we are saddled with a great deal of responsibility. The decisions that we make, on how fast or how far we go on the space program, are going to decide whether we retain our present position as compared to Russia, whether we advance it, or whether we slow it down.

This is the responsibility that we have. That is why we have so many questions.

Mr. MALLAN. I appreciate that.

Mr. KARTH. This is why we feel so very strongly. Probably in some cases you might feel we are being derogatory in our line of questioning, but we have some real decisions to make. I appreciate what Mr. Daddario said, that if we assume that this is gospel, this story, "The Big Red Lie," that you have written, if we should assume that and then recommend to this Congress that we proceed on less of a crash-type basis—we don't have a crash program as it is, but even to a lesser degree—then of course we would be more seriously in a position of less responsibility for the leadership which I think we are looked to for, from the military preparedness standpoint.

This is a real decision we have to make.

Mr. MALLAN. I understand that. I would like to make it very clear that I do not think our space research program should be cut back one iota. I also feel that the United States must, has to, maintain world leadership in scientific research of all kinds. I think more emphasis should even be placed upon this so that we would have a great big backlog of knowledge. We have a big backlog now, but we need to make use of every possible creative scientist or even practical technician.

Mr. KARTH. The point I make is that if Congress should accept "The Big Red Lie" as the gospel, this Congress would cut appropriations for military preparedness, military expenditures, in my opinion quite considerably from what it is today. If this Congress were convinced of your story, irrespective of my desires or the desires of this committee, I think that the expenditures for military preparedness would be cut drastically.

Mr. MALLAN. You say military preparedness as if that is a special thing.

Mr. KARTH. The field of air, the field of ground, the field of almost anything.

Mr. MALLAN. How about scientific preparedness, which in a way includes military preparedness?

Mr. KARTH. You know as well as I do that we have made our greatest scientific advancements generally during times of war.

Mr. MALLAN. That is true, but aren't we a more intelligent people than that?

Mr. DAIGH. Could I make one more short statement?

The cold war, I think, should be a sufficient stimulant because there has never been any secret that Russia despises us, will wipe us off the face of the map at the first opportunity and gives us demonstrations at all times. We can be secure that we are ahead of them, but I can't myself justify or see the justification of a democracy making a war effort under false information.

The kind of false information can drive a dictatorship—it is characteristic of a dictatorship—that the mass of people are given wrong information and driven to frenzy heights with wrong information.

This is not characteristic of a democracy and should be no part of it. Our health will grow better with our panic. There is a do-good in panic. If you will forgive me, I can see it reflected in the questions you ask, that all of you men have been very thoroughly indoctrinated, and with good reason, I am sure, with the idea that every Russian is 15-feet tall, has four pairs of hands and carries a missile under each arm. This is not the truth.

Mr. KARTH. It is not true as far as I am concerned. Everybody else can speak for himself, but this isn't the way I view Russia.

Mr. DADDARIO. Mr. Daigh, if I might add a comment, I think you have come to the same conclusion that some of the members of this committee have, when you say that the scientists of Russia have not been at all affected by the Russian system, that they could not conceivably tell a lie. This committee, almost all of its members, have a great deal of information concerning the Russians and some of them have had practical training.

I feel we should go along with the President of the United States in this statement, and taking into consideration all of the intelligence that he had at his beck and call and would support his position, because I think that he has a great variety of sources, a great deal of ability to check, much beyond and above that of Mr. Mallan.

When we try to come to conclusions here, as we have, we are doing it only so that we can come to a proper conclusion and we have asked questions not only of Mr. Mallan, but of the scientists who come before this committee in the same manner.

Mr. DAIGH. Could I add I think it would be most unfortunate——

Mr. KING (presiding). The Chair will have to intervene at this point.

Gentlemen, we appreciate your views. We have another witness whose testimony we want to be certain to do full justice to; so I will have to intervene at this point. We are very grateful, Mr. Mallan, and Mr. Daigh, for your appearing. The fact that our cross questioning has been somewhat relentless must not obscure the fact that we are very grateful for your presence. You have made a real contribution. I want to thank you for it.

Could we call Dr. Alan H. Shapley of the Bureau of Standards, Boulder, Colo., station. I understand, Dr. Shapley, you come here from Boulder for purposes of being with us. We are grateful for your effort also. If you would give your full name and your position to the reporter.

Dr. SHAPLEY. My name is Alan H. Shapley. I am Assistant Chief of the Radio Propagation Physics Division of the National Bureau of Standards, Boulder Laboratories, at Boulder, Colo.

Mr. KING. Because of the investigative nature of this proceeding, we would like to administer the oath to you, if you would raise your right hand, please.

You do solemnly swear that the testimony which you are about to give before this subcommittee in the matters under consideration will be the truth, the whole truth, and nothing but the truth, so help you God?

Dr. SHAPLEY. I do.

Mr. KING. Be seated, please.

STATEMENT OF ALAN H. SHAPLEY, ASSISTANT CHIEF, RADIO PROPAGATION PHYSICS DIVISION OF THE NATIONAL BUREAU OF STANDARDS, BOULDER LABORATORIES, BOULDER, COLO.

Mr. KING. I don't know, Dr. Shapley, whether you have a formal prepared statement or whether you are speaking impromptu; but, such as it is, you are invited now to present your statement.

Dr. SHAPLEY. Mr. Chairman, I was invited to appear here, I understood, not because I would have any fresh information to provide, but to answer or respond to questions which might be put to me, particularly dealing with radio astronomy.

I am very happy to help in any way that I can to the extent of my competence and experience. I should make it clear that the National Bureau of Standards at its Boulder Laboratories are engaged in several programs involving radio astronomy. We patrol the sun and measure the radio emissions from the sun and record the solar outbursts. Also we make ionospheric observations by observing radio stars, natural radio stars, and also make ionospheric studies by observing beacon transmitters carried in earth satellites.

We are not a tracking station as such at the Boulder Laboratories, but rather a scientific station. With that explanation of the degree of the experience of the Bureau of Standards in this problem, I am willing to try to answer any questions which you might have.

Mr. KING. Dr. Shapley, our able counsel has prepared a series of questions that might possibly get at this. You understand, of course, that our common interest here is whether or not a lunik was launched and whether it performed more or less as the Russians claimed it to have performed. It is upon that question that we are trying to throw light.

Would you feel more at home to just talk to that, or should I go through these questions which have been prepared?

Dr. SHAPLEY. I think it would be better if you went through the questions, Mr. Chairman, because I myself, nor our laboratory—we haven't had the job of tracking or of determining whether such vehicles actually exist. So I have no new information to give to you.

Mr. KING. Fine. With that in mind then, I will proceed to relay a few questions that Mr. Beresford has prepared, which I think may elicit some of the information that you are prepared to give.

Are you able to say at what time the Soviet moon rocket was launched? Do you happen to remember that?

Dr. SHAPLEY. I am afraid I cannot testify as to that.

Mr. KING. Just for the sake of the record, the time on that was 5 p.m., Greenwich time, January 2, 1959. If you will check your notes, you will find that to be correct.

Assuming that burnout occurred at approximately 5 minutes after launching and the vehicle had then just reached escape velocity, at what time would it pass closest to the moon? Maybe I am throwing some tough ones at you. I am not sure I even understand that after reading it.

Dr. SHAPLEY. Perhaps if I had been given these questions on the telephone last Monday, which I am afraid was the first time I heard of this hearing, and of this business-----

Mr. KING. I understand. I might say we have received testimony as to the time when the Goldstone tracking station received its signals. The purpose of the questions, of course, was to see whether your independent computations would coincide with their alleged reporting. Of course, it is unfair to throw this at you without giving you a chance to compute it.

Mr. BERESFORD. Perhaps Dr. Shapley could submit his computation for the record at a later date.

Dr. SHAPLEY. I would be glad to do that.

Mr. KING. Fine. No sooner had I read the questions than I realized it would be impossible to expect an answer without computation, but if you could compute that and submit that for the record, we would be most grateful.

Do you want me to repeat the question?

Dr. SHAPLEY. Could you, please?

Mr. KING. Assuming that burnout occurred approximately 5 minutes after launching, and the vehicle had then just reached escape velocity, at what time would it pass closest to the moon, remembering that the launching time was 5 p.m. Greenwich time, January 2, 1959.

Dr. SHAPLEY. Five p.m.?

Mr. KING. Five p.m.

Dr. SHAPLEY. 1700 hours?

Mr. KING. Is that question sufficiently detailed for you to come up with an answer?

Dr. SHAPLEY. Yes; it is.

(The information requested is as follows:)

If the vehicle were launched at just the velocity of escape and along the shortest path to the vicinity of the moon, and if we neglect the moon's gravitational attraction, the vehicle would reach the moon's distance in 49.3 hours. This comes from the formula:

$$t=\frac{2}{3\times V\times r^{1/2}}\left(D^{3/2}-r^{3/2}\right)=177{,}000\text{ seconds}=49.3\text{ hours}$$

where t=travel time
V=velocity of escape=11.2 km./sec.
r=radius of earth=6370 km.
D=distance to moon=384,000 km.

Thus if the launching were under these conditions and were at 1700 hours universal time on January 2, the vehicle would pass closest to the moon at about 1800 hours universal time on January 4.

However, the travel time is very sensitive to the exact initial velocity. If the initial velocity were 1 percent greater than the velocity of escape, a more complicated calculation shows that the travel time would be 38.3 hours, and if 10 percent greater the travel time would be only 18.2 hours. If the initial velocity were 1 percent less than the velocity of escape, the vehicle would never reach the moon's distance; if four-fifths of 1 percent less, it would reach the moon's distance with zero velocity in 116 hours or 4.83 days.

Taking into account the moon's gravitational attraction renders the problem more complicated, but because the mass of the moon is so small (one-eightieth that of the earth) the effect is felt only when the vehicle is in the close neighborhood of the moon. As far as travel time is concerned, the effect would be small except in the cases where the initial velocity was less than the velocity of escape.

The altitude of burnout does not significantly affect these calculations except for the time required for vehicle to reach that altitude.

Mr. KING. There may be some other questions of the same category. If you are not prepared to answer or if it will take time for you to compute, of course we will understand. What was the position of the Moon at that time and the position of Jupiter?

Dr. SHAPLEY. Mr. Chairman, Mr. Beresford did indicate there might be a question along that line, so I computed the separation of the Moon and Jupiter, if that would be sufficient. I can, if you like, from the "American Ephemeris Nautical Almanac" give you the actual positions. Do I understand you are really interested in the difference of positions?

Mr. KING. You understand, Doctor, what we are driving at. It has been alleged that what was thought to be lunik signals were actually Jupiter signals, and what have you, and we are trying to establish the basis for either settling or repudiating such alternate hypotheses.

Dr. SHAPLEY. Mr. Chairman, as seen from the center of the earth, the distance between the center of the Moon and Jupiter, I have computed for January 4 at zero hours, January 5 at zero hours, and so on. I can provide this table for you if you would like. At January 4 at zero hours, Greenwich time, the separation was 21°23′, and Jupiter was southeast of the Moon at that time.

On January 5 the separation was 7° 24′.

(The information requested is as follows:)

Angular distance between Moon and Jupiter (geocentric parallax not included)

1959 Jan.	4.0	21° 23′	Jupiter southeast of Moon.
	5.0	7° 24′	Do.
	5.5	2° 02′	Jupiter south of Moon.
	6.0	7° 25′	Jupiter southwest of Moon.

Mr. KING. Pardon me, Doctor. Does that mean that on the celestial horizon the Moon and Jupiter would be 7° apart.

Dr. SHAPLEY. That is right. The angle from the center of the earth to these two bodies would be 7° on January 5.

I should emphasize that this is computed for an observer at the center of the Earth. The actual difference for a person on the surface of the earth would be changed somewhat because what is called the geocentric parallax of the Earth amounts to as much as, I believe, about 1°. So these figures would have to be modified, depending upon where the observatory was that was in question.

Mr. KARTH. Mr. Chairman.

Mr. KING. Mr. Karth.

Mr. KARTH. On the basis of what you know insofar as the location of the tracking station, insofar as the information that was released from Russia on the proposed course of the lunik, and insofar as you know as a matter of fact where Jupiter was, is it possible that these signals that were received at Goldstone could have come from Jupiter?

Dr. SHAPLEY. There are many factors involved. If you will allow me to recall that the acceptance angle of the antenna at Goldstone is, I believe, somewhere between 2° and 4° and if I can assume that, then I would say that it was not possible to confuse the two anytime on January 4.

Mr. KARTH. Anyone understanding these fundamentals would not have suggested that these signals came from Jupiter; is that correct?

Dr. SHAPLEY. That is correct. Not only for this reason, but because of the frequency on which the observation was made. Jupiter's atmosphere radiates in the radio frequencies rather strongly, as you have heard. It radiates strongly, though, in a very restricted part of the frequency band, somewhere between 18 and 30 megacycles. Jupiter is a very bright, a very powerful radio source in that frequency range.

Mr. KARTH. Could these signals have come from any other natural source—natural source, not manmade?

Dr. SHAPLEY. There are no other natural sources I know of, distant sources, in the direction mentioned. The apparent source seen by the Goldstone dish moved through the sky at a different rate than the rotation of the earth, or seemed to.

Mr. KARTH. Is it possible these signals might have been received as they bounced off an airplane or two crossing the path?

Dr. SHAPLEY. I have not done calculations along those lines but just from a general calculation of radio-field strengths this is a most unlikely circumstance. It would also take very peculiar positioning of radio transmitters on the surface of the Earth. Either that, or some rather curious faults in the equipment, or both. I think it is very unlikely.

Mr. KARTH. Are you discharging my question with the thought that it is impossible, or that there is a very, very vague hypothetical possibility?

Dr. SHAPLEY. I guess we scientists do not like to deal in absolutes. I think in all of science it is always best to speak in terms of probabilities. Here the probability, I think, would be exceedingly small.

Mr. KARTH. Let me ask it this way, then, Doctor. From what you know about science, can you discount my question as one that is impossible or one that is very vaguely hypothetical?

Dr. SHAPLEY. It is certainly more nearly impossible than vaguely hypothetical.

Mr. KING. Dr. Shapley, how close would this lunik have to approach the Moon before it would be able to measure the Moon's radioactivity and magnetic field? Would you have an opinion on that?

Dr. SHAPLEY. It would of course depend upon the sensitivity of the instruments that were carried; I would estimate radioactivity—what was the other?

Mr. KING. The magnetic field.

Dr. SHAPLEY. The Moon's magnetic field. With presently available instruments, I would estimate—and this is a rough estimate—I would estimate it would have to be two or three of that order, two or three lunar diameters away.

Mr. KING. A lunar diameter is in the order of——

Dr. SHAPLEY. About 2,000 miles. I would like to correct that if I am badly off.

Mr. KING. I believe that is approximately correct. Moving on, what is the approximate bearing error of radio telescopes and tracking equipment such as that of Goldstone, Fort Monmouth, Jodrell Bank?

Dr. SHAPLEY. I do not know. The errors are different for each installation. I know generally about the Jodrell Bank and the Goldstone installation, but unless you know the details of the instrument, it would be impossible to put it in terms of bearing errors. You can compute the angular width of the beam, the acceptance beam of the instruments. I believe for an 80-foot dish at 200 megacycles, it is somewhere between 2° and 4°. That is about all an outsider can do in estimating the errors of such instruments.

Mr. KING. Would what you have just said be a fair generalization as to the bearing error of a tracking station of first magnitude such as you have mentioned?

Dr. SHAPLEY. Oh, no. The final errors would be much, much smaller than that.

Mr. KING. Much smaller?

Dr. SHAPLEY. Much smaller.

Mr. KING. But you are not in a position to be more specific without more specific information?

Dr. SHAPLEY. That is correct.

Mr. KING. This next one again you may have to compute. If so, you can get it into the record.

How long a chord is subtended at a distance of 250,000 miles by an angle of 1 degree?

Dr. SHAPLEY. If you do not mind, I will supply that for the record. I can work it out in a few moments.

(The information requested is as follows:)

A chord of 4,365 miles subtends an angle of 1 degree at a distance of 250,000 miles.

Mr. KING. I will just hand you this list of questions, but there are some you can comment on here. For that reason I will continue to read them.

What do you understand by the term "tracking" and, in connection with that, is it correct that either range information or cross-bearings are required? How many cross-bearings do you consider necessary for reasonable scientific accuracy?

Dr. SHAPLEY. Mr. Chairman, that is a question on tracking and I was brought here more as a radio astronomer. It would be much more of an appropriate question for someone like Mr. Mengel or Dr. Pickering, the trackers.

Mr. KING. I have two more in that same category. You may want to take the same position on them. Is it necessary for the points of intersection obtained from cross-bearings to coincide? How large an error would you consider allowable at a distance of about 250,000 miles? Is your position the same on those two?

Dr. SHAPLEY. It is the same; yes.

Mr. KING. Is it possible to derive location, speed, or distance from radio signals? Would any assumptions be involved as to initial position, time of launch, and so forth?

Dr. SHAPLEY. Location, speed, and distance?

48438—59——13

Mr. KING. Yes. Is it possible to derive location, speed, or distance from radio signals?

Dr. SHAPLEY. Yes. It is possible certainly to get speed. Again this is not strictly a radio astronomy question. One can, though, get the speed by the Doppler effect on radio signals. To get that, though, one does need to know very accurately the frequency of the transmitter carried in the vehicle.

Basically, radio signals of just ordinary beacon transmitters in a vehicle will give you the direction of the vehicle.

Mr. KING. That Doppler effect is the effect whereby sound rises in pitch as the body approaches——

Dr. SHAPLEY. That is correct; yes. The same thing also applies in light and in radio.

Mr. KING. The next question you partially answered: What is Doppler? Is it possible with Doppler to estimate the location, speed, or distance of a signal source?

Dr. SHAPLEY As I say, the speed can be measured. I believe you do need other information in order to get the location. You need some additional information. Again I must comment that I am not an expert in this particular field.

Mr. KING. Would any assumptions be involved as to absolute transmission frequency?

Dr. SHAPLEY. You would have to know that or have to assume it correctly in order to use the Doppler principle to get the speed; and you would have to moreover assume that the transmitter frequency stayed constant over the period of your observation.

Mr. KING. Could Doppler be caused by accidental or intentional variations in transmitter frequency?

Dr. SHAPLEY. Yes; it could be. It could be quite difficult to closely reproduce the Doppler effect, but it could be done.

Mr. KING. In your opinion would it be possible to simulate the signals emitted by a rocket launched into outer space? How, and is there any reliable method of determining whether such signals are simulated or authentic?

Dr. SHAPLEY. I think it would be a very difficult project to do such a simulation. I almost was going to say as difficult as that of launching the rocket.

You, I imagine, are referring to an observation of the kind that has been reported to us by the Goldstone place. The only way that a terrestrial signal at these very high frequencies can appear to come down from the sky would be by a mechanism like what is called ionospheric forward scatter. This is where the radio signal is not reflected, but is scattered by the F layer—the high layer in the ionosphere.

To do this, you would have to have a transmitter at an appropriate distance away. And in the case of the signals that were observed near the moon for a total period of several hours, you would then have to move your transmitter in space with the change in the angle of the downcoming signal. This would be almost so fantastic that it couldn't be imagined.

Further, the greatest angle at which you can get such a scattered signal coming down from the sky would be less than 10 degrees. I believe it has been discussed here that that angle was considered to be more than 10 degrees and varied through the night.

So I do not see any way of simulating the signal by means of ionospheric forward scatter. I have not been able to think of any other schemes for such simulation that would not also require moving a terrestrial transmitter rather speedily, moving its distance on the ground rather quickly.

So I do not see any plausible way of simulating such a transmission.

Mr. King. Can you suggest any explanation for the failure of the Jodrell Bank radio-telescope to detect lunik signals on 183.6 megacycles, even though it was aimed directly at the moon for several hours before and after the announced time of closest approach?

Dr. Shapley. I have no information on this so I cannot explain it. I will comment—as I think one of the Congressmen did earlier this afternoon—that there is a great difference between carrying out an experiment which has been planned for a long time, where the equipment has been ready and tested and you know what to look for, when to look for it, and all that—there is a great difference between that and the improvisations which are necessarily in order on an unexpected launching of this sort.

Whereas the receiver may be tuned very precisely to the new frequency, it is probably a different receiver than is normally used because you have to have variable tuning in that case; whereas in, say, tracking a satellite or a space vehicle where you have had lots of time to prepare, then you would have a crystal controlled receiver right on the frequency.

By having to change to a different receiver, you probably lose the optimization of the receiving elements so your sensitivity undoubtedly suffers.

I do not know whether these were factors in the Jodrell Bank case, but I know that these factors have come up whenever one has to improvise a scientific experiment.

Mr. King. Dr. Shapley, I have come to the end of my questions. I am not a technical man in this field, which fact is very obvious. Mr. Beresford on my right is knowledgeable in this field.

I am going to make the suggestion, therefore, that we recess for 5 minutes to give you and Mr. Beresford further opportunity to talk this over to see whether there are further questions which you think should be asked to elicit further information that will contribute to our knowledge of this matter. You have come a long way, and we want to be certain that we have every bit of information that you feel is necessary or that you can contribute. Even though we have lost most of our audience, I want to assure you that your testimony goes into the record. It will be read and digested thoroughly by members of this subcommittee.

So, without objection, I will declare a recess for 5 minutes to enable you and Mr. Beresford to collaborate on that matter.

(A short recess was taken.)

Mr. King. The subcommittee will resume its session.

I should like to introduce into the record, Mr. Reporter, material handed to me by Mr. Daigh, which consists of the seven reasons referred to by Mr. Mallan as to why the Russians could not—what is it they could not do?

Mr. Daigh. They could not have an ICBM or have fired a lunik.

Mr. King. The material will be incorporated into the record.

(The documents referred to are as follows:)

Explanations

1. No Guidance System

The Russians claim lunik passed within 4,650 miles of the moon. In terms of outer-space distances, that's a very near miss—a superb piece of celestial marksmanship. To achieve that kind of accuracy, you need an extremely precise, reliable system for starting your rocket in the right direction and holding it on course.

There are two kinds of guidance system that can give you this kind of accuracy. One is radio control from the ground. The other is inertial navigation.

They could not have used ground radio control because, if they had, the signals would have been picked up by free world tracking stations. The type of equipment used in remote radio control generates very powerful signals. These radio signals would have been heard all over the world by our monitor stations just as surely as we'd hear a gunshot fired in the back of this room.

The Russians could not have used inertial guidance, either. The reason: simply that they have none to use. Inertial guidance depends on a complex set of miniature instruments which sense and control the rocket's motion in flight. To build such a system, you need miniaturization, high-precision engineering, and a very good but very small computer. These things are beyond the Russians' capabilities.

I became convinced of this—convinced that the Russians couldn't build the kind of inertial guidance system needed—simply by studying Russian engineering. In addition, I asked Russian scientists about their experiments in inertial guidance.

Many scientists I asked had never heard of inertial guidance. At the Soviet Academy of Sciences, the fountainhead of all scientific activity in Russia, I talked with Profs. Yury Pobedonostev and Kiril Stanyukovich, both key scientists in the Soviet space effort. I asked, point blank, whether Russia had developed an inertial guidance system.

Russia would like us to think she's far ahead of us in the conquest of space; and I know that, if these men chose to lie to me, they'd lie in the direction of exaggerating Soviet progress and Soviet cleverness.

But they didn't lie. This was Stanyukovich's answer to my question:

"Some work in this field has been done, but it is all theoretical so far."

Of course he didn't know I would interpret his answer, and answers of others, as proof Russia does not have the ICBM, and later, no luniks.

2. No Miniaturization

It requires a tremendous amount of power to hoist even an ounce of weight out into space. At the same time, a rocket is a very complex device, packed with equipment. To get all the necessary equipment in without adding bulk, you must miniaturize—build your components as small as possible.

Miniaturization has been one of the most important avenues of technological advance in the United States over the past 10 years. We've developed such mighty midgets as the transistor, which can handle many jobs of the much larger vacuum tube. We've even gone a step smaller, into microminiaturization—with such fantastic results as a radio transmitter the size of a matchbox.

There has been no such development in Russia. The Russians hardly seem interested in miniaturization. They engineer with a heavy hand. They admire big things.

This fact was pointed up by the sputniks. Premier Khrushchev scoffed at the U.S. Vanguard satellite because it was so small in comparison. He called it the "Grapefruitnik." What he didn't say was that the Vanguard is a much more useful scientific instrument that any of the sputniks. It was launched into orbit with such precision that it will stay up there for at least 2,000 years, long after Sputnik III follows I and II to destruction. Its tiny batteries and tiny transmitter are still sending loud, clear signals—while those of Sputnik III are fading.

The Russians built their sputniks big because they didn't know how to make them small. In the words of Col. John Stapp, president of the American Rocket Society: "Those who can't build watches must build grandfather clocks."

This grandfather clock construction is acceptable in earth satellites, where only a relatively small amount of equipment will do. It doesn't take much instrumentation, comparatively speaking, to lob a chunk of metal into a crude orbit about the earth. But to send a rocket to the moon requires a lot of sophisticated, accurate equipment.

At the Polytechnic Museum in Moscow, where the pride of Soviet science is displayed, I saw no evidence of miniaturization. There were some vacuum tubes that were labeled "small," but were actually big compared to tubes made in the West.

The Russians are known to make transistors. This isn't much credit to their engineering skill, however. Soviet transistors are exact copies of General Electric models.

Not even the top scientists I talked with seemed interested in miniaturization. I carried with me a miniature tape-recorder—a little device that often seems to surprise and delight Western scientists who see it for the first time. Of all 38 Russian scientists who saw it, only two were even vaguely interested. The rest apparently thought it unimportant because it was small.

Even if the instrument package on lunik weighed the amazing 790 pounds they claimed, the Soviets could not have fitted all the instruments they said were on board, without miniaturization.

8. NO MINIATURE COMPUTER

When a spacecraft is in flight, it's subjected to a complicated interplay of powerful forces, all trying to turn it off course. If it is to head accurately toward its target, it must have on board a small, fast, clever electronic computer. This computer must be light so that the rocket can haul it aloft, and it must be capable of making literally thousands of calculations and decisions every second.

Such a device can only come out of a very sophisticated computer technology. Russia has none. She is years behind the United States in this science.

I studied computers carefully when I was in Russia. I looked at the latest textbooks on the subject, saw their computers and pictures of them, talked with scientists about computers. I learned that the most up-to-date computers produced in Russia are equivalent to machines that IBM and Remington Rand were mass-producing 6 or 7 years ago.

In the fast-moving computer world, 7 years are a long time. Whole generations of computers have come and gone in that period.

At this year's annual meeting of the Institute of Radio Engineers, members were told that Russian scientists have only now begun thinking theoretically about cybernetics and information theory. These are two immensely valuable fields of study. They long ago passed the theoretical stage in this country, and have since helped to boost the speed, capacity, and versatility of our computers.

The Russians would like us to believe they have advanced computers like ours. I know of at least two cases where they hoaxed us on computers as they hoaxed us on lunik.

In one case, the Russian Army paper Red Star carried a picture of a very modern-looking computer installation, in conjunction with an article about Soviet progress in the field. Remington-Rand was much interested in the picture. It had been stolen directly from a Remington-Rand advertisement in Time magazine, according to Gen. Leslie Groves, former head of Manhattan project and now an official of Remington-Rand. Only one change had been made: Remington-Rand's name had been blacked out of the picture.

Another Soviet magazine, a leading science publication, carried a feature story about Prof. Gleb Chebotarev, director of the Institute of Theoretical Astronomy at Leningrad. Chebotarev is one of Russia's top mathematical astronomers. The article spoke breathlessly of the "huge, automatic computing machines" he used in his work. I wasn't particularly impressed by this, for I knew that American mathematical astronomers have used large computers for years—wouldn't think of working without them, in fact.

I went to visit Professor Chebotarev, and I asked him about the "huge, automatic" computers he used.

His answer: "I am not acquainted with these machines. All my calculations are made by another professor with the help of small electrical calculators."

Gentlemen, I ask you, is there any conceivable propaganda reason for one lone scientist to deny the official published boasts of the Soviet in a field where Russia has consistently made extravagant claims.

4. NO ATOMIC CLOCK

Space navigation is an art of split-second timing. You must control the exact moment when each piece of equipment goes into action. If your timing is the slightest bit off, your whole project may end in failure.

As an example of the kind of timing required, take the question of when to shut off a rocket's motor. After the initial burning period, the rocket coasts the rest of the way to its target or orbit. If the motor is shut off too soon or too late, the rocket will be traveling too slow or too fast—and won't do what it was supposed to do.

According to the National Aeronautics and Space Administration, a rocket headed for Venus must be traveling at about 37,000 feet a second when its motor shuts down. If it misses the required speed by as little as 6 inches a second, it will miss Venus.

That takes real timing, and there's only one kind of clock that can split seconds accurately enough. This is the atomic clock. It takes its timing from the pulses or vibrations inside atoms, and it is accurate to a millionth of a second or better.

The Russians have no such clock. I asked many scientists about it—among them Prof. Vladimir Tsesevich, a top meteor research man, and Prof. Dimitri Rozhkovsky, an important astrophysicist. Tsesevich and Rozhkovsky both said that they understood American scientists were working on an atomic clock, but admitted that Russia hasn't yet started working on any such device.

The Russians don't even have good crystal clocks, which are also high-accuracy devices—though they don't compare with atomic clocks. At the space-tracking stations I saw in Russia, the crystal clocks in use were American-made. And even these weren't working as accurately as they were supposed to, for the Russians had them crudely hooked up with mechanical time-printing devices. I knew those clocks were built to be accurate to a thousandth of a second or better. The Russians told me, proudly, that theirs were accurate up to a three-hundredth of a second.

To launch a rocket to the moon, you must be able to cut time into thousandths and even millionths of a second. The Russians simply haven't learned how to do it.

I have recently computed that if only one of many guidance components in the alleged lunik was off time by one five-hundredths of a second, the rocket would miss the moon by about 50,000 miles. To have come within 4,650 miles of the moon, as claimed, the Russian timing would have had to have been accurate within one five-thousandths of a second.

5. NO OPTICAL TRACKING

Before you can hope to make improvements fast in space technology, you need to know exactly how your rockets behave in flight. That's why accurate visual and photographic tracking are essential. Observations of exhaust flames, for instance, give you valuable information about the functioning of a rocket's engine. Without photographic records mistakes cannot be corrected and superior rockets constructed.

The Russians have no optical system to do this job.

One of my key interviews in Russia was with Alla Massevich, who heads Russia's optical space-tracking network. She is vice president of the Astronomical Council of the Soviet Academy of Sciences, and she is one of Russia's leading astrophysicists.

She took me up on the roof of the Sternberg Astronomical Institute, where her headquarters are, and proudly showed me a typical Soviet optical tracking station. The major piece of equipment was an ordinary Red Air Force aerial reconnaissance camera, of a type that you can buy here through war-surplus outlets for less than $80.

Miss Massevich told me: "We have 25 tracking stations like this at universities and observatories all across Russia."

I subsequently visited six of them, and all were the same. All used the antique—by our standards—camera.

Compared to optical tracking in the United States, the Russian system is laughably inadequate. In our tracking network, we use specially designed cameras built by Perkin-Elmer Corp. at a cost of $100,000 each. We don't merely set them up at existing universities or observatories, but spot them at carefully picked locations for the best possible tracking effectiveness.

Throughout my stay in Russia, I looked and questioned hard to turn up evidence of a sophisticated optical tracking system. I found none. One professor was working on plans for a camera something like our Perkin-Elmer instruments, but he did not expect to develop a finished camera for several years.

Aside from his work, the entire Soviet optical tracking network had the air of a haphazard scramble to throw something together with whatever equipment happened to be lying around. There was no evidence of the kind of careful advance planning that has gone into American tracking efforts.

Could the Soviets have hidden high-grade tracking equipment and entered a conspiracy to show me outmoded, jerry-built equipment?

I will concede that the Russians are tricky. They lie, but it is impossible to get from them a statement they might believe would be interpreted as unfavorable to Russia.

It is well known to the free world that Alla Massevich heads Russia's optical space-tracking network.

On past performance any statement she might give for publication in the free world that would make Russia look bad would be for her to invite a one-way ticket to Siberia.

Russia is still bombarding the free world with boasts of her alleged technological superiority in optical tracking.

It is not conceivable that one reporter, one little writer of books, would be deliberately misled in contradiction to the longstanding Communist propaganda line of Soviet superiority—a line in use everywhere today.

Is it conceivable that better equipment, which, if they had it, would weigh tons, would be moved out of sight in all 25 of the known tracking stations in Russia and antique jerry-built tracking cameras substituted on concrete bases? Incidentally, if there is anyone that naive in my audience, the concrete bases were weathered and had obviously been there some time.

This situation I personally inspected in 6 of the 25 known stations.

6. NO RADIO TRACKING

Radio tracking goes hand in hand with optical tracking. You use it to see how your rockets are behaving in flight, to detect errors in their design, to bring back information from them on conditions in outer space. Each rocket thus supplies data that you can use to make your next rocket better.

Russian radio-tracking equipment is primitive. At several major observatories I visited, including the Crimean Astrophysical Observatory, the most important in all Russia, the radio equipment my Soviet hosts showed me was so backward that it embarrassed me.

American scientists have grumbled often that the Russians never release any new or useful data from their sputniks. I say that the Russians have nothing exciting to release—no effective radio equipment to gather the data.

I'd like to cite two revealing incidents.

Some 3 months after the Russians launched their first satellite, Sputnik I, Soviet tracking equipment lost it. An official announcement from Moscow said that the satellite had burned to nothing in the atmosphere.

This told us much about the effectiveness of Russian equipment. Five days after the Russians lost their own satellite, astronomers at Ohio State University's radio observatory were still tracking fragments of the satellite that remained in orbit.

Obviously, the Russians were using a much more crude, much less sensitive kind of equipment than Ohio State's.

Another incident: Three weeks after Sputnik I's launching, Jodrell Bank Radio Observatory, Manchester, England, got a frantic cable from Russia.

Would Jodrell Bank be so kind as to locate the lost sputnik? Sputnik's radio had gone dead and Russia couldn't find her propaganda rocket by any means she had.

Jodrell Bank obligingly made a quick radar fix in conjunction with their radio telescope and sent the required information to the helpless Soviet scientists.

7. NO SERVOS

If a lunik is to be under the control of a miniature computer, there must also be miniature servomechanisms—mechanical devices to carry out the computer's commands. The servos are the computer's hands.

The Russians are not skilled in building servomechanisms—particularly miniature ones. In factory automation, which uses computers and servos in the same way as a rocket, Soviet engineers are years behind us. The Soviet magazine, Automation, publishes wide-eyed reports about developments that American engineers considered old hat during World War II.

For instance, the Russians think they're pretty clever when they can get a lathe to shape a piece of metal automatically. A man sets up the job, switches on the lathe, steps back, and watches the tool do the rest of the work. Here in the United States we were doing that kind of thing back in the 1940's. We're now working on eliminating the man entirely—making the entire factory automatic.

The Russians haven't had to pay as much attention to automation because human labor is much cheaper there than here. They haven't had to put nearly as much thought into computers and servos. As a result, they simply haven't developed our level of skill in designing and building these devices.

In recent years, there has been a good deal of breathless admiration in the Soviet press over automatic signal systems that are being installed on Russian railroads. In the United States such systems were introduced at the turn of the century.

Or take another field, aviation. On all big American airliners, wing flaps and other control surfaces are moved by servomechanisms, not human muscle. It's much like power steering in an automobile. American planes have used this system since before World War II. But Soviet engineers still haven't caught on to the idea. Even in the supposedly ultramodern TU-104, the famed Soviet jetliner, there are no servos for the control surfaces. To make a turn in flight, the pilot and copilot have to throw their combined brute force against a lever.

No nation that backward in building automatic equipment could have any hope of getting a lunik close to the moon.

Mr. KING. Dr. Shapley, I believe during the recess you were able to discuss one or two matters and focus your attention on some in which you are competent to testify which will throw light on this general question.

Rather than asking you detailed questions, I think it is in order simply for you to discuss the matter.

Dr. SHAPLEY. Mr. Chairman, I realize my comments on the radio emanations from Jupiter were perhaps incomplete. Jupiter in this lower part of the frequency spectrum between, say, 18 and 30 megacycles is a very, very bright object. The strength of the radiations from Jupiter above 30 or 40 megacycles drops off very, very fast. In fact, this kind of radio radiation from Jupiter has not been observed at frequencies greater than 50 megacycles.

It has been looked for on 85 megacycles and 153 megacycles with very powerful radio telescopes without any positive results. In fact, the biggest radio telescopes in the world have been put onto this problem.

So I think that the possibility of radiation being received from Jupiter on 186 megacycles is out of the question in view of these other observations.

To make the statement fully complete, I should say that at much, much higher frequencies, one can observe thermal radiation from Jupiter up around 3,000 megacycles and at still higher frequencies. But this thermal radiation has not even been observed at frequencies in the neighborhood of 183 megacycles.

I believe earlier I got diverted or interrupted when I was talking about Jupiter's spectrum.

A second thing I might comment on is the possibility that ionized clouds in nearby space could bend radiation and therefore give you a false idea of the location of the source. It is true that the ionosphere does bend, does refract, radio waves. In fact, that is one of the important studies being made by the Bureau of Standards at the present time.

The amount of this bending, though, in the ionosphere is of the order of 1 degree or less and would, I don't think, not be a significant amount for what we have in mind.

As regards ionized clouds out beyond the earth's atmosphere, we don't have direct evidence of dense ionized clouds in the vicinity of the earth; and clouds, to give a bending of as much as 1 degree would certainly have other properties that would have made them observable.

So I believe that it is most unlikely that bending by ionized clouds in the vicinity of the earth could complicate our present problem.

It has been suggested that I comment on the weights of the sputniks. I am probably not the best person to do that. You should have a celestial mechanician to do this. But it is possible by studying the orbit of these sputniks and by judging from how bright they are optically, and with some knowledge of how dense the high atmosphere is, to make an estimate as to the weights. There is one other thing you have to know, and that is how the object is tumbling as it goes around in its orbit.

I understand that, even though our knowledge of all of these factors is rough, that the weights given by the Russians for the sputniks are of the right order of magnitude to explain the observed orbits and the changes of the orbits.

This doesn't prove it, but the results of these observations and calculations are consistent with the figures given by the Russians.

Mr. KING. That concludes your testimony, does it, Dr. Shapley?

Dr. SHAPLEY. Yes, it does.

Mr. KING. May I reiterate my expressions of gratitude on behalf of the subcommittee for your appearing here, for your very valuable contribution.

Mr. DAIGH. Mr. King, may I ask one question of the witness—very short?

Mr. KING. Yes, you may.

Mr. DAIGH. I would like to ask whether or not it is a fact that for the first 3 days of this week Dr. Shapley was in New York City serving on a committee in the U.N. with Dr. Dryden?

Mr. KING. Would you like to answer that?

Dr. SHAPLEY. Yes.

Mr. DAIGH. Thank you.

Mr. KING. Anything further? Mr. Mallan, you were going to furnish for the record two names which you prefer doing other than in open session.

Mr. MALLAN. That is correct.

Mr. KING. You don't have to do it right now.

Mr. MALLAN. Only for the sake of the people themselves. For all I know, by now they are——

Mr. KING. We will respect your desires. After the session is adjourned perhaps you can furnish that information to Mr. Beresford.

Mr. MALLAN. I will be glad to do it.

Mr. KING. Fine. I appreciate that. Did someone else have a question?

Dr. SHAPLEY. I wanted to amplify my answer to the previous questions, to say that the chairman of this subcommittee is also a member of the subcommittee in New York on which I have been assisting.

Mr. KING. Thank you. Anything further?

Mr. MALLAN. I would like to ask Dr. Shapley merely one question. Would it be possible, since the signals from Jupiter between the 18 and 30 megacycle range are so strong, that they might not have harmonics that fall within the 183.6 megacycle range?

Dr. SHAPLEY. No signals from Jupiter have been observed, as I say, on 85 and 153 megacycles when searched for with the very large telescopes. I think this is clear evidence that there are not strong harmonics in the radiations from Jupiter.

Mr. MALLAN. May I add that both Dr. Frank Drake of the National Radio Astronomy Observatory and Dr. John Kraus, director of the Ohio State Radio Observatory, both of whom have been for the past several years observing Jupiter, both answered in the affirmative, one slightly less enthusiastically than the other, that harmonics are quite possible from that range of frequencies of Jupiter.

It was Dr. Frank Drake who said we must certainly expect that to be so, and Dr. John Kraus said it is most likely.

Mr. KING. Do you care to comment on that, Dr. Shapley?

Dr. SHAPLEY. No; I don't think so.

Mr. KING. Your statement stands in the record.

Dr. SHAPLEY. My statement stands.

Mr. KING. Any further comments?

Mr. MALLAN. One other thing about ionized gas clouds in near space. Dr. Shapley says it is unlikely that these clouds could have refracted or bent the signals from Jupiter. On the other hand, Dr. Kress says there is a possibility, but little is known about these clouds, their density or their very existence. So this leaves that question kind of open, I think.

Then one final point and that is the weights of the sputniks. I can't agree that we could measure the weights on the basis of so little data in terms of the tumbling and the density of the ionosphere which shifts very often, Dr. Shapley, depending upon outbursts from the sun and various other sources—a sudden large influx of ultraviolet radiation which might change the entire characteristics of a layer and even create a new layer such as the D layer, isn't that so?

Unless you know specifically what the conditions are at a certain place and time, how can you be sure of your calculations? That is my question.

Mr. KING. You are not required to answer, but if you care to, you may.

Dr. SHAPLEY. To the best of my knowledge, no deviations of this magnitude in the positions of any extraterrestial sources have been observed. I think that enough work has been done on it so that if this were an important feature that they would have been observed. They certainly have been looked for.

Mr. MALLEN. I said what was the final question, but I do have one more.

Mr. KING. Mr. Mallan, we will allow one more question, but we will have to bring this to a close.

Mr. MALLAN. Thank you very much. Dr. Shapley, as a scientist, can you say that nothing is impossible, or that there are limits to science? Simply because something hasn't been observed before—for example, Newton didn't observe the physical universe around him as completely as Dr. Einstein did. So for 200 years people accepted these limitations. This slowed up the fastest moving science in the world today, physics.

So would you say on that basis that it is impossible that certain special conditions could not exist that would contradict a whole background of investigation and theory that is accepted as standard?

Mr. KING. Again you may comment, if you wish.

Dr. SHAPLEY. I don't think a comment is in order at this time.

Mr. KING. You have made your point and we are grateful for it. Anything further? There being no further business before the subcommittee, we will stand adjourned.

(Whereupon at 5:10 p.m. the subcommittee adjourned, to reconvene at 10 p.m., Friday, May 29, 1959.)

SOVIET SPACE TECHNOLOGY

FRIDAY, MAY 29, 1959

House of Representatives,
Committee on Science and Astronautics,
Special Subcommittee on Lunik Probe,
Washington, D.C.

The subcommittee met, pursuant to adjournment, at 10:10 a.m., Hon. Victor L. Anfuso (chairman of the subcommittee) presiding, in room 356, Old House Office Building.

Mr. Anfuso. This meeting will come to order.

I am going to recall Mr. Mallan. Please come up, Mr. Mallan.

STATEMENT OF LLOYD MALLAN, AUTHOR OF "THE BIG RED LIE"

Mr. Anfuso. Mr. Mallan, I want to inform you that you have already been sworn yesterday and that you are still under oath. You understand?

Mr. Mallan. I understand that.

Mr. Anfuso. Mr. Mallan, can you possibly tell this committee where exactly your father was born?

Mr. Mallan. It was somewhere in Poland, a portion of Poland that is now a part of the U.S.S.R.

Mr. Anfuso. That is the best information you can give us?

Mr. Mallan. Well, I know—I don't know, but I am fairly sure that my father left Europe before the U.S.S.R. ever took over this part of Poland.

Mr. Anfuso. You have no idea as to the name of the town?

Mr. Mallan. No; I don't.

Mr. Anfuso. Whether or not the town is still in existence?

Mr. Mallan. No; I don't. In fact, as I recall, my father left Europe before there was even a revolution in Russia. He left when he was around 18, I believe.

Mr. Anfuso. He left when he was 18?

Mr. Mallan. I think so.

Mr. Anfuso. How many years ago was that?

Mr. Mallan. Well, he died about 8 years ago or 9 years ago.

Mr. Anfuso. How old was he then?

Mr. Mallan. He was around in his seventies then. I am just talking approximately, but this was my impression.

Mr. Anfuso. It was before 1900 then, is that it?

Mr. Mallan. It must have been.

Mr. Anfuso. About 1905?

Mr. Mallan. I can't say. I am sorry.

Approved For Release 2004/08/31 : CIA-RDP63T00245R000100290011-7

Mr. ANFUSO. I want the record to show that Mr. Mallan has appeared before this subcommittee voluntarily—as a matter of fact, he has asked to appear before this subcommittee and that he has thus far been very cooperative with the subcommittee. There isn't any question about that.

Mr. KING. No question.

Mr. ANFUSO. Mr. Mallan, when you received this passport to go to Europe, could you tell us how your name was spelled on it?

Mr. MALLAN. Well, as I recall it was spelled the way I normally spell my name.

Mr. ANFUSO. Lloyd Mallan—M-a-l-l-a-n?

Mr. MALLAN. That's right.

Mr. ANFUSO. Is there a possibility it might have been spelled in some other way?

Mr. MALLAN. There is no possibility I would spell it any other way on my application for a passport.

Mr. ANFUSO. Somebody could have made a mistake and spelled it differently?

Mr. MALLAN. I don't know.

Mr. ANFUSO. Did you ever see written on any passport or any other document the name spelled M-a-l-l-e-n?

Mr. MALLAN. You mean on a document that was mine?

Mr. ANFUSO. Yes; passport or otherwise.

Mr. MALLAN. I wish I could answer that.

Mr. ANFUSO. Did you ever see it, I mean?

Mr. MALLAN. I have seen people write my name M-a-l-l-a-m—I mean, you know, on a piece of paper or something. I have seen them make mistakes, if that is what you mean.

Mr. ANFUSO. Yes. Anyway, to the best of your recollection, the passport that you received was spelled, as far as you know, M-a-l-l-a-n?

Mr. MALLAN. That's right. As a matter of fact, I remember that when I was applying for it, the woman who typed out the form for me was typing it just as I spelled it to her, at least on the form.

Mr. ANFUSO. Do you understand Russian?

Mr. MALLAN. No. I mean, after having been in Russia, I can catch about three words out of the entire language—five words out of the entire language.

Mr. ANFUSO. Could you read Russian?

Mr. MALLAN. Now I can make out words like "academy" and "science." That is about it.

Mr. ANFUSO. Just to go back a bit to the individuals that you say contacted you before you left this country to go to Europe: When you returned to the United States did you see those individuals again?

Mr. MALLAN. I may have bumped into one, but I never made a point of seeing them.

Mr. ANFUSO. Did any of them contact you upon your return?

Mr. MALLAN. No.

Mr. ANFUSO. Not to your recollection?

Mr. MALLAN. No. In fact, I am fairly sure none of them did.

Mr. ANFUSO. Have any of them contacted you since that time?

Mr. MALLAN. No.

Mr. ANFUSO. You say one of them died?

Mr. MALLAN. One of them was killed in Spain.

Mr. ANFUSO. Was killed in Spain?

Mr. MALLAN. Yes.

Mr. ANFUSO. Was this around the same time you were there?

Mr. MALLAN. I believe this was a little later.

Mr. ANFUSO. In other words, he followed you there and came afterward?

Mr. MALLAN. That is right.

Mr. ANFUSO. And you learned while you were there that he had been killed?

Mr. MALLAN. No; I didn't learn that until after I had come back. I read about it.

Mr. ANFUSO. You read about it?

Mr. MALLAN. Yes.

Mr. ANFUSO. His name appeared in the newspapers?

Mr. MALLAN. I read about it in a book. In fact, it was the other day while we were trying to check on the battalion and the brigade relationship.

Mr. ANFUSO. Could you tell us the name of the book, if you will, or do you have any reason not to tell it?

Mr. MALLAN. No; I have no reason not to tell it.

Mr. ANFUSO. Would you tell us the name of the book?

Mr. MALLAN. I really don't remember. It was an excerpt that was obtained by one of the researchers at True magazine, who went to the public library and just patiently went through all the material there.

Mr. ANFUSO. You did not read the name in any newspaper?

Mr. MALLAN. I may have.

Mr. ANFUSO. You may have?

Mr. MALLAN. I don't recall.

Mr. ANFUSO. Mr. Mallan, when you returned from Russia you went through Germany, didn't you?

Mr. MALLAN. That is right.

Mr. ANFUSO. Where in Germany?

Mr. MALLAN. I came back by way of East Berlin and then stayed in West Germany, mainly at West Baden.

Mr. ANFUSO. West Baden in West Germany?

Mr. MALLAN. That's right.

Mr. ANFUSO. Was it your intention to stay there 5 months?

Mr. MALLAN. No; it wasn't.

Mr. ANFUSO. Were you interrogated by any U.S. officials?

Mr. MALLAN. Yes; I was.

Mr. ANFUSO. Can you state by whom?

Mr. MALLAN. Yes; by Air Technical Intelligence.

Mr. ANFUSO. Anyone else?

Mr. MALLAN. By all of the specialists at Air Technical Intelligence.

Mr. ANFUSO. How long were you interrogated?

Mr. MALLAN. Well, off and on during the entire 4½ or 5 months.

Mr. ANFUSO. Were you asked to stay there until they completed the interrogation?

Mr. MALLAN. They asked me to volunteer the information. In fact, I volunteered to tell them whatever I could.

Mr. ANFUSO. Did you also volunteer to stay there as long as they wanted you to?

Mr. MALLAN. Yes; I was willing to.

Mr. ANFUSO. Is that why you stayed that long?

Mr. MALLAN. That is one of the main reasons.

Mr. ANFUSO. What did they ask you?

Mr. MALLAN. Each time they would bring a new type of specialist over to talk with me. They covered the entire field, every field that I was familiar with—aviation, air defense, aircraft engines, electronics, optical equipment—just about everything—missiles.

Mr. ANFUSO. More or less did they cover the field that this committee has already covered and the material which you reported in True magazine?

Mr. MALLAN. Yes—well, except for the lunik which was launched allegedly after I got back to the States.

Mr. ANFUSO. But all of your visit there, the places you went to, and the data that you acquired, they inquired about?

Mr. MALLAN. That is right.

Mr. ANFUSO. And you at no time withheld any information?

Mr. MALLAN. I even gave them my notes and drawings for copying, and my tape recordings, and I have a letter from them thanking me for this, if you wish it for the record.

Mr. ANFUSO. Yes.

Mr. MALLAN. I will get it for you. It is back in my briefcase.

Mr. ANFUSO. That letter will be inserted at this point in the record.

(The letter is as follows:)

HEADQUARTERS,
U.S. AIR FORCES IN EUROPE,
New York, N.Y.

To Whom It May Concern:

During their stay in Wiesbaden, Germany, Lloyd and Rose Mallan spent considerable time with Government personnel at the request of the U.S. Air Forces, Europe; this time was considered by the Air Force very necessary and well spent.

It is hoped that the services rendered by Air Force personnel, such as providing tape transcriptions, etc., provided some measure of compensation for the Mallan's extremely willing cooperation.

E. T. DOOLEY,
ATI Directorate,
Headquarters, USAFE.

Mr. ANFUSO. Any further questions of Mr. Mallan?

We have a few other witnesses. Does anybody else want to question Mr. Mallan on anything that I brought out?

All right, thank you very much, Mr. Mallan.

Mr. MALLAN. You are welcome.

Mr. ANFUSO. You will stand by, won't you, Mr. Mallan?

Mr. MALLAN. Yes, sir, I will.

Mr. ANFUSO. Mr. Slavin.

Mr. SLAVIN. Mr. Chairman, I would like to have Dr. Curtis with me, if I may.

Mr. ANFUSO. Yes, Mr. Slavin. Would you please state for the record your full name and title and also the full name and title of Dr. Curtis?

Mr. SLAVIN. I am Robert M. Slavin, director of Project Space Track. This is Dr. Harold Curtis, the Chief of the Operations Branch of Project Space Track.

Mr. ANFUSO. Mr. Slavin, I understand you have a prepared statement which is rather short—a page and a half. Would you mind reading that statement, and then we will ask you some questions.

Mr. SLAVIN. I will, sir.

Mr. ANFUSO. Go right ahead.

STATEMENT OF ROBERT M. SLAVIN, DIRECTOR, PROJECT SPACE TRACK, AIR FORCE CAMBRIDGE RESEARCH CENTER; ACCOMPANIED BY DR. HAROLD CURTIS, CHIEF OF OPERATIONS BRANCH, PROJECT SPACE TRACK

Mr. SLAVIN. I appreciate this opportunity to appear before you today to discuss the Russian lunik probe into space.

Space Track has the Defense Department responsibility for maintaining a catalog and for issuing predictions of future position on all artificial satellites and space probes. In doing this work a control center at Bedford, Mass., serves as a focal point for the collection and analysis of data.

The observations are made by a widespread network, including observers at Air Force test ranges, cooperating universities and observatories, and elements of other agencies such as Army, Navy, and NASA with which we maintain an interchange of information. The observations from this network normally include radio and optical measurements from a large variety of instrumentation.

The processing of these data and the computation of future positions are carried out at the Space Track Control Center at Air Force Cambridge Research Center, from which bulletins are distributed to the observing network.

Specifically, on January 2, 1959, following the Russian announcement of the launch of a lunar probe, the control center issued a bulletin to all those stations which were part of the Space Track network. The bulletin requested monitoring and, if possible, tracking on the announced transmission frequencies.

During the next 30-odd hours, many observations of these frequencies were reported to the control center. Of these, only the reports from Goldstone and Stanford Research Institute contained tracking data, the remainder being reception of signals at the announced frequencies.

The achievement of the Goldstone and Stanford groups in obtaining tracking data in the short time available was outstanding. The successful tracking of a vehicle when the time of launch, the transmitted frequencies, and the planned trajectory are unknown quantities until after the launch, is very difficult.

The best that can be done under these circumstances is to obtain fragmentary tracking data and to evaluate these data using our own knowledge of such trajectories. In this case, this conclusion is that the similarity of our data to the Russian announcements confirms the claim that a lunar probe was launched and did in fact reach the vicinity of the moon.

Mr. Chairman, following any questions that might be asked at this point, we would like to present in closed session a short briefing prepared on this matter by the Air Force Intelligence people. I think you will find the material quite pertinent to the inquiries.

48438—59——14

Mr. ANFUSO. Mr. Slavin, we are going to ask you some questions in this open session and then restrict the major part of our questioning to the closed session. If you think you ought to reserve the right to fully explain some of these questions I ask you in closed session, you may do that, too.

Mr. SLAVIN. Thank you.

Mr. ANFUSO. Would it be technically possible to simulate the lunik signals on 20 megacycles by terrestrial transmissions?

Mr. SLAVIN. I would like to ask Dr. Curtis to comment on that, if I may.

Mr. ANFUSO. Surely.

Dr. CURTIS. My first comment on this is that almost anything is technically possible. It is a question of how much trouble you want to go it. The 20-megacycle signals are, by and large—I will make that stronger, with one exception—only receptions of a signal. Only one of these contain any sort of tracking data. Simply, then, the process was turning on a radio receiver—normally, just an ordinary communications receiver—and listening at the frequencies announced by the Russians.

The signals heard—the three frequencies at 20 megacycles were time shared—on three separate transmitting frequencies which were time shared. Some of the observers reported this time sharing.

Mr. ANFUSO. Just explain what you mean by "time sharing"?

Dr. CURTIS. There were three frequencies: 19.997, 19.995, and 19.993 megacycles. This is the transmitted frequency of the signal. I have forgotten the exact time sharing, but one was on, say 0.997, then 0.995 then 0.993. Some of the stations—I have forgotten again which ones; it is available in our data—did observe this time sharing of the signals.

Mr. ANFUSO. What you are saying, then, Doctor, is that it is possible to simulate the signals, but one would have to go through special efforts to do that?

Dr. CURTIS. I think so. The other thing that can be noted about these signals is that the ones which were reported to us occurred when the moon and presumably the lunar probe was overhead. Only one or two of the signals are when the moon and the lunar probe would have been on the other side of the earth.

Mr. ANFUSO. When the moon is where—on the other side?

Dr. CURTIS. On the other side of the earth from the observer.

Mr. ANFUSO. Could these signals have been simulated by airborne transmissions?

Dr. CURTIS. They certainly did not originate on any known satellite.

Mr. ANFUSO. On any known satellite?

Dr. CURTIS. From any known satellite. The only satellite transmitting near this frequency was the Russian Sputnik III. This was tracked by most of these agencies during that day. I think the question "could it be done from any other sort of airborne transmitter" is again a question of to what effort do you want to go.

Mr. ANFUSO. Could it be simulated, for example, by the kind of aircraft the Russians are known to have?

Dr. CURTIS. I can't answer that. I don't know much about Russian aircraft.

Mr. ANFUSO. Could it be simulated by any American known aircraft?

Dr. CURTIS. Aircraft and ground simulation is about the same problem, I think.

Mr. ANFUSO. It is about the same.

Dr. CURTIS. Yes; there is not much difference.

Mr. ANFUSO. Could it be simulated by bouncing signals off the moon or an artificial earth satellite?

Dr. CURTIS. I have no knowledge of the possibility of bouncing 20-megacycle signals off the moon, sir. To my knowledge it hasn't been done.

Mr. ANFUSO. That is highly improbable?

Dr. CURTIS. I don't know what anybody has ever tried.

Mr. ANFUSO. Is it possible to estimate the distance or speed of a signal source from signals such as were obtained from Goldstone?

Dr. CURTIS. Now we are changing from the signal at 20 megacycles to the signal at 183.6 megacycles. The Goldstone antenna is a pure listening antenna. That is what that antenna is. It doesn't transmit anything, it just listens. It sits there and essentially looks like a parabola, and as the signals come from space, the antenna focuses those signals into the pickup for the radio receiver. Then by moving the antenna you can locate approximately—and this depends on the type of signal with which you are dealing—you can locate the source, and then the antenna points at the source. Then you read off the two angles which define that point in space.

Now then it does not measure distance in any way. It only measures two angles. To get any knowledge of speed from this angle measurement requires also a knowledge of distance. You can get it that way. If the signal——

Mr. ANFUSO. You can't get the speed unless you have the distance?

Dr. CURTIS. That's right; yes.

Mr. ANFUSO. And you can't get the distance?

Dr. CURTIS. You cannot get the distance from this simple measurement made in that way. If the signal had had an observable doppler shift—this is a shift in transmitted frequency, due to the speed of the vehicle, then if you knew what the real transmitted frequency was, then you could get a measure of the speed away from the antenna or toward the antenna from the change in frequency of the transmission.

However, in this case we certainly did not know the exact Russian frequency. The problem of putting together high grade, low-noise electronics in that time is formidable, and the result of this—and this is from the Goldstone people—they would not have expected to have seen a doppler shift due to the instability of their own equipment.

So the answer, in all, to the question is there was no measure of speed.

I would like to say something more about this, but not in response to a direct question. I have here a plot of the data obtained at Goldstone. The black dots at the bottom are the position of the moon on the fourth of January at about 1200 Greenwich mean time—1200 to 1400 Greenwich mean time. The Goldstone observations are all centered about 1300.

The next series of circles is the position as reported from the Russian press release of the lunik. That is this series of dots along here. The little crosses above are the positions of the declination—this is one of the astronomical measurements of position, one angle, the position of the probe as measured by Goldstone, the position at which the antenna was when it heard this signal or this series of points here.

Also of interest here, this circle is the apparent size on this diagram of the moon. In other words, in this one angle it looked as if, according to the Russian announcement, the probe had just missed the periphery of the moon.

Let's go to the second one. This is the other angle that would be measured. This is essentially the longtiude measurement. On this we plotted the same thing. Here is the position of the moon, these three points, the black ones. The circles again, the little circles are the positions from the Russian announcements of where the lunar probe was, and again the crosses are the positions measured at the Goldstone antenna.

Here the agreement looks much better mainly because the scale is compressed. The actual disagreement is about the same as on the other plot but we had more information to get on the plot.

It is this agreement that convinces us that the Russian announcements are correct.

Mr. ANFUSO. What assumptions would be involved in estimating speed or distance? What assumptions which you have to have?

Dr. CURTIS. From this data?

Mr. ANFUSO. From this data.

Dr. CURTIS. You would need to assume where it was, how far away it was.

Mr. ANFUSO. How could you find out?

Dr. CURTIS. Except from the Russian announcement, we could not.

Mr. ANFUSO. The only way we could determine that would be from the Russian announcement; is that right?

Dr. CURTIS. That is right. We have no independent measurement of distance.

Mr. ANFUSO. We have no independent measurement of distance or speed, I would say.

Dr. CURTIS. Yes; that is correct.

Mr. SLAVIN. May I add one point to that. These plots that Dr. Curtis has just showed you indicate, of course, that this thing is moving across the sky with the speed of the moon. If we take anything in close, such as an airplane, this is practically impossible that it would move that slowly. So therefore the assumption is that it is far out.

Mr. ANFUSO. It is far out?

Mr. SLAVIN. Yes; then the correspondence of these data with the Russian announcements which results in this plotted trajectory indicates a great deal of agreement with these announcements.

Mr. ANFUSO. Could the Rusians have confused our people by these announcements?

Dr. CURTIS. I think it is the other way round, sir. On the basis of the Rusian announcements, the Goldstone people were able to point the antenna and confirm the Russian announcements.

Mr. SLAVIN. I would like to speak to this point of simulation. When we take the 20-megacycle signals about which the questions were asked, these are technically possible. It is technically possible to simulate the effects that we had up to a certain point. But when you take the whole picture of all the signals, then the simulation would become as big a job probably as launching a shot to the moon.

Mr. KARTH. Mr. Chairman, may I ask a question?

Mr. ANFUSO. Yes, Mr. Karth.

Mr. KARTH. Yesterday, Doctor, it was the opinion of Mr. Mallan that it would be a fairly simple matter to simulate the data that we received by artificial means. Do you then disagree with that contention?

Mr. SLAVIN. I think so, when you look at these plots that you have just seen. I think the ability to get this much correspondence becomes a pretty complicated affair.

Mr. KARTH. There was some talk yesterday—not talk, this was the testimony of Mr. Mallan—as I recall, that there is a very grave possibility that radio signals from the ground could have bounced off of airplanes traveling in this vicinity which would have assimilated the results of your finding. Is that possible?

Mr. SLAVIN. Dr. Curtis would like to comment on that.

Dr. CURTIS. I have a feeling that the fact that only the Goldstone site observed this may be significant in the answer to this question. If the source were low, then the power should have been available and the transmission should have been hearable at other places, in which case triangulation of two places looking at the signal would have occurred—in which case they would point definitely to a low source.

Mr. KARTH. Wasn't the power available in other tracking stations around the country so as to have been able to pick up the signals that you picked up on the lunik, if there was a lunik?

Dr. CURTIS. I think the 183.6 megacycle signal was only picked up at Goldstone because only at Goldstone were they able to build an excellent enough receiver in the day that they had and only at Goldstone did they have—not only at Goldstone, but they had an excellent big receiving antenna——

Mr. ANFUSO. Will you yield there just a minute, Mr. Karth? Testimony was given by Mr. Mallan, and he produced a Russian magazine which indicated that hams throughout Russia were able to pick up these signals. If hams throughout Russia were able to pick up these signals, why is it that only Goldstone picked them up?

Dr. CURTIS. I imagine, sir, these were signals at 20 megacycles.

Mr. ANFUSO. What is the difference? Explain that to us.

Dr. CURTIS. Twenty megacycles is in the normal communication band. Equipment exists, it has been well developed over the years for communications purposes. Actually the communications receivers are almost as good as are available. At 183.6 megacycles this is only a frequency which has been exploited for television and not in any very great way by the hams or by communications people, for long distance transmission.

So that is a telemetry frequency essentially which is used in this way only scientifically.

Mr. ANFUSO. If the Russians say that Russian hams picked up these signals, do you think that they are doing a little bit of boasting?

Dr. CURTIS. I wouldn't doubt the claim at all because when the vehicle is launched in the first few hours it is rather close. This is sort of an order of magnitude statement, I think, the first time any place in the United States had an opportunity to hear this signal, the lunar probe was already in the vicinity of 60,000 miles away. This is a tremendous distance for radio transmission.

Mr. ANFUSO. In other words, Russian hams only heard the first signals?

Dr. CURTIS. I would think so, yes.

Mr. ANFUSO. If they claim they heard further signals they would be exaggerating?

Dr. CURTIS. The 20-megacycle signals we report here could have been heard by hams.

Mr. ANFUSO. Some of them.

Dr. CURTIS. They were weak signals. They were in the noise.

Mr. ANFUSO. How close would lunik have to come to the moon in order to measure its radioactivity and magnetic field?

Dr. CURTIS. I would think very close.

Mr. ANFUSO. How close in miles?

Mr. SLAVIN. The best measurement of course would be made by going right in to the surface. Depending on the sensitivity of the experiment, I have seen numbers on the magnetic field which indicates that you would not get much of a measurement unless you were within a few hundred miles.

Mr. ANFUSO. Mr. Newell said 100 miles, and Mr. Shapley said 4,000 to 6,000 miles. What is your comment?

Mr. SLAVIN. My comment is a few hundred miles.

Mr. ANFUSO. A few hundred?

Mr. SLAVIN. Depending on the sensitivity of your instruments, and this is using techniques which are known, a few hundred miles.

Mr. ANFUSO. All right, Mr. Karth.

Mr. KARTH. Doctor, if 20 megacycles is so easy to pick up, why didn't we have some of our tracking stations tuned in at 20 megacycles to track this thing during the course of its first thousand miles?

Mr. SLAVIN. It wasn't available to our tracking stations until it was some 60,000 miles away from the surface of the earth.

Mr. KARTH. How would that have an effect?

Dr. CURTIS. It is below the horizon.

Mr. SLAVIN. It is below the horizon. It is on the other side of the world during the initial portion of its flight.

Mr. KARTH. So it was impractical for us to tune very quickly to that type of a frequency because, for all practical purposes, we would have received no intelligent information; is that correct?

Mr. SLAVIN. We did have some recordings of the 20-megacycle signals at a later time.

Mr. ANFUSO. Mr. Slavin, I understand you talked to Mr. Mallan; is that correct?

Mr. SLAVIN. That is correct, sir.

Mr. ANFUSO. Mr. Mallan claims that you told him, in answer to his direct question about the possibilities of the Russians faking a signal from somewhere in the Earth, the following: that Soviet submarines could be waiting somewhere out at sea in international waters, that it could wait for the Moon to rise above the horizon and beam a radio signal toward the United States.

Did you tell him this?

Mr. SLAVIN. I believe I probably did. I would like to comment on that.

Mr. ANFUSO. Please.

Mr. SLAVIN. The date of this discussion—I don't remember precisely when it was, it was sometime early in January following the——

Mr. ANFUSO. About January 13 he says.

Mr. SLAVIN. January 13? Yes. At that time all the data had not been examined thoroughly. We had not reached any definite conclusion. He raised the possibility of this thing being a hoax, and, of course, we always like to keep an open mind on these things. This was a possibility.

Mr. ANFUSO. As Dr. Curtis says, if anybody wanted to do it, I suppose you could have a Soviet submarine do exactly what Mr. Mallan said, if the Russians wanted to do such a thing. Is that correct?

Mr. SLAVIN. I believe that the effort that would have to be mounted in order to accomplish this, when you consider all the data, would be rather monumental.

Mr. ANFUSO. It is possible, but the probabilities aren't there; is that what your testimony is?

Mr. SLAVIN. Yes, sir.

Mr. WOLF. Could you give us some idea of the physical effort involved if they were going to simulate this thing?

Mr. SLAVIN. If you were to simulate the motion, as these plots have showed, the motion of a signal at the rate of the moon across the sky, the only way I can really think of doing this is to get a transmitter out at that particular point and move it at that rate. This thing then becomes a lunar probe.

Mr. WOLF. How long have you been in this work?

Mr. SLAVIN. Since 1948 I have been associated with the Air Force's upper atmosphere rocket program, and since 1957 associated with the satellite tracking field.

Mr. WOLF. This has been your full enterprise for almost 2½ years?

Mr. SLAVIN. Yes, sir.

Mr. ANFUSO. One more question I have, and then I will submit you to other Members of Congress and counsel, too. If counsel have any questions, they may be permitted to ask them.

Mr. Mallan also said, when he asked you how accurately the Soviet scientists could determine the velocities and distances they announced for their lunik as it passed beyond the moon, you said emphatically, to use Mr. Mallan's words—he quotes you as saying:

> When I read the report in the newspapers that their vehicle had passed by the moon at 4,650 miles, I snorted. How the hell could they possibly determine that? With the most accurate optical and electronic equipment, the determination could be accurate only to within 1 degree, and this is being generous. That would make it not within an accuracy error range of 50 or 100 or even 1,000 miles, but within about 5,000 miles of possible error alone. This is only considering the moon alone. In the case of vehicles moving beyond the moon, the error to be considered would be even greater.

Did you say something like that, or substantially——

Mr. SLAVIN. I feel there may be some overemphasis in some respects there, possibly on my part, too. I know the answer to the question was given in the sense that the last significant digit, the 50, I believe it was, miles, is rather ridiculous, because I don't really believe that anyone knew the position to this accuracy.

When you follow the whole trajectory with tracking which you have prepared for months in advance for a launch of this sort, when you follow the whole trajectory by this method, you could probably get closer than my 1-degree estimate.

If you were making a single measurement out of this spot, you are probably lucky if you get within this range.

Mr. ANFUSO. Let me ask you this, Mr. Slavin. Could we have independently determined that this vehicle passed the moon at approximately 4,650 miles?

Mr. SLAVIN. I doubt if we would ever have gotten down to the 50. I doubt probably if we would have got much closer to the 100's. I would like Dr. Curtis to speak on this.

Mr. ANFUSO. Suppose we calculate this to the 100's and were to accept this figure of 4,650 miles as the figure. We must accept that figure from the Russians. Is that it?

Mr. SLAVIN. Might I ask Dr. Curtis to comment on this matter of accuracy?

Mr. ANFUSO. Yes, indeed.

Dr. CURTIS. I think in general that in the problem of tracking space probes you know—with your tracking information available at launch—more than you will ever know again about a space probe. The launch data then is the most accurate data. On the basis of the launch data you can prepare the look angles, the direction in which to aim these big antennas. The big antennas are nowhere near as precise as the launch data.

Mr. ANFUSO. In other words, you have to have the launch data before you can accurately determine a distance of this nature; is that it?

Dr. CURTIS. I think that the way you would narrow down the distance to get the ultimate accuracy out of a tracking system is to use the launch data as a basis of computation, and then correct this by using the later tracking data available from this radio telescope sort of antenna.

So the figure, I think, for the distance of closest passage to the moon would come mainly from the launch data and computation. I think it would be impossible to make a measurement of the vehicle at that distance which would indicate the distance of closest approach to that accuracy.

Mr. ANFUSO. Dr. Curtis, are you saying that in order for us to determine the exact or near location of a satellite, or an intercontinental missile, that we would have to have the launch data?

Mr. SLAVIN. I think Dr. Curtis is referring to the far space probe, rather than the near one.

Dr. CURTIS. I am talking about real distant objects.

Mr. WOLF. Mr. Chairman, could I have him define "launch data," exactly what we are talking about?

Dr. CURTIS. Launch data would be mainly radar data of the vehicle during launch, determining a series of positions and a series of velocities very accurately.

Mr. ANFUSO. That would be confined to Russian territory. That is what you wanted to know, isn't it?

Dr. CURTIS. This would be confined to the first 5 minutes of flight, something like this.

Mr. ANFUSO. You say this does not hold now with intercontinental missiles. We could, from our shores, for example, by means of what we are developing, the Nike-Zeus, determine an intercontinental missile location.

Mr. SLAVIN. The basic difference here, of course, is the matter of triangulation, in that when you are out in space at 200,000 miles away, the earth is only 8,000 miles in diameter and this is a very small fraction of it. When you are up in the satellite altitude of 300 miles, or ICBM altitude, then your base line on which you can make measurements is greater than the altitude which you are trying to measure. So you can get more accuracy out of it that way.

Mr. ANFUSO. I just want to satisfy my own mind and the minds of the members here. I am sure that we are on the right track, as far as intercontinental missiles.

Dr. CURTIS. Oh, yes. Let me make a statement about unclassified radars in operation. The best of these is the Millstone radar of the Lincoln Laboratory. This is the only radar which can routinely look and see a satellite at great distances.

The error there, stated error, of the Millstone system is something like small fractions of a degree, one- or two-tenths of a degree of angle error, and a range error of something like 5 miles.

Mr. ANFUSO. Five miles?

Dr. CURTIS. Yes.

Mr. ANFUSO. That is the best?

Dr. CURTIS. With that radar.

Mr. ANFUSO. We will ask you some questions in closed sessions, because I would like to have that pursued. That is a very important point.

Any other questions? Mr. King?

Mr. KING. Just one little matter. You showed us this plot a minute ago in which the three positions of the moon were plotted out, and then you showed us the trajectory as computed of this lunik, and it coincided with the position of the moon.

The question in my mind is: the Russians would know that, too, and they would know the exact position of the moon at different periods of time. Suppose they had just manufactured this whole story, but had so done it that their story would coincide with their foreknowledge as to where the moon would be at certain times. Couldn't they have——

Dr. CURTIS. Oh, yes; they certainly knew the position of the moon, and if they fired a lunik they aimed at that position. The significant thing is there was a transmission, an object transmitting a signal on 183.6 megacycles from that position.

Mr. KING. It coincided——

Dr. CURTIS. It coincided with where the Russians said their lunik was and it coincides very closely to the position of the moon also. One thing that perhaps should be clarified. These signals, I said, were clustered around 1300 Greenwich time. The Russians said the lunar probe passed closest to the moon at about 0300 Greenwich time. So in listening to this, the signal apparently is coming from beyond the moon, but at about the same direction as the moon.

Mr. KING. From what you say it is not significant if it coincided with what the Russians said it should be, because they could manufacture that. It is significant that it did coincide with where you knew the moon was. Am I correct?

Dr. CURTIS. No, sir; no. You misunderstood my emphasis here. Let's leave the question of the moon out of this for the moment; where the moon is. The Russians announced a trajectory for the lunik probe, which goes through a certain number of points in space. This signal was heard coming from those points in space at the time at which the Russians said the lunar probe was there. The moon does not enter into that question.

Mr. KING. In other words, they announced beforehand, or simultaneously with launching, that we launched it at such-and-such a time the course will be such-and-such?

Dr. CURTIS. Yes, sir.

Mr. KING. We focus our apparatus up into the air, either at the spot where it is supposed to be, according to their trajectory, and lo and behold, it is there, it coincides?

Dr. CURTIS. Yes, sir.

Mr. KING. Then in addition we have the second point that it also follows the moon which it was supposed to do?

Dr. CURTIS. Yes, sir.

Mr. KING. You know where the moon is, you know where your signals are coming from, the two coincide?

Dr. CURTIS. Yes, sir.

Mr. KING. Of course, it could be argued, I suppose, that that just proves these signals may have come from the moon itself.

Mr. SLAVIN. They are too far away, sir.

Dr. CURTIS. They are a little bit too far away to come from the moon itself. The error is significant, the deviation in hour angle, particularly, essentially a longtitude measurement.

Mr. KING. So you have ruled out the possibility of its being a reflection from the moon?

Dr. CURTIS. It is very, very improbable it would be a reflection from the moon.

Mr. KING. That is all I have.

Mr. KARTH. Mr. Chairman, if I could just pursue this a moment. Because of the position of the moon at the time when they said they were firing their lunik, the trajectory which they chose to give you the path they said the lunik would follow almost had to be the path; didn't it?

Mr. SLAVIN. There were several possibilities. They picked one of them.

Mr. KARTH. How many possibilities were there?

Mr. SLAVIN. Actually an infinite number, but there is a certain spread in time of travel, for example. They picked a 32-hour travel. There was a possibility of a 48, 52, something of this nature.

Mr. KARTH. Explain that. What do you mean?

Mr. SLAVIN. In other words, 32 hours from launch to impact.

Mr. KARTH. Is that the shortest time that was possible?

Mr. SLAVIN. I am not certain whether it is. I believe it is.

Mr. KARTH. How about the doctor? What do you say? Is that the shortest time possible?

Dr. CURTIS. Possibilitywise I don't want to comment. This coincides actually quite closely——

Mr. KARTH. In the best of your opinion would that be——

Dr. CURTIS. It coincides quite closely with the attempts of Pioneer III and IV, the manner in which NASA attempted those flights.

Mr. KARTH. But they have chosen a different trajectory?

Dr. CURTIS. They could have. For instance, the Air Force lunar probe attempts used something like a 60-hour flight time. It depends upon the parameters of the launching vehicle, on the excellence of the guidance system you use, the total characteristics of your equipment.

Mr. KARTH. Considering the greater hour figure that you might choose for the vehicle to fly, as the number of hours increases, does the difficulty of making this a successful shot increase proportionately?

Dr. CURTIS. I don't know.

Mr. SLAVIN. I don't know. There are so many parameters that vary here that it is hard to reach a definite conclusion on this.

Mr. ANFUSO. Mr. Wolf.

Mr. WOLF. I just wondered if Dr. Curtis would like to put in the record the length of time you have been involved in this business.

Dr. CURTIS. I first was assigned to Project Space Track in February of 1958.

Mr. WOLF. What is your background before that?

Dr. CURTIS. I have been at the Air Force Cambridge Research Center in Atmospheric Physics since 19—.

Mr. WOLF. You have quite a background in this particular business.

Dr. CURTIS. In the field of general instrumentation and high atmospheric physics I have a fair background, yes.

Mr. ANFUSO. Any further questions?

Mr. KARTH. Just one, Mr. Chairman. Mr. Mallan suggested that there was a possibility these signals could have been coming from Jupiter. Do you agree with that?

Dr. CURTIS. One moment. On the basis of some data obtained from the JPL, the Goldstone people, I would say it is very, very improbable that the signals did originate on Jupiter. It is in the wrong direction—not tremendously, but outside the limits of the possible error of the apparatus.

Mr. DAIGH. Mr. Anfuso, would you recognize me for a minute?

Mr. ANFUSO. Yes.

Mr. DAIGH. It is so fortunate to have these two gentlemen here and listen to their interesting testimony. I wonder if we couldn't perhaps be further illuminating to your committee by permitting Mr. Mallan to ask these gentlemen one or two significant questions?

Mr. ANFUSO. It is very unusual that such a thing is permitted, but I will allow one or two short questions.

Mr. DAIGH. I thank you very much.

Mr. ANFUSO. Mr. Mallan.

Mr. MALLAN. Dr. Curtis——

Mr. ANFUSO. In the interest of getting at the truth; True magazine is supposed to tell the truth. Go ahead.

Mr. MALLAN. Before, when you were talking about the 20 megacycle range of frequencies, you mentioned that after 60,000 miles it would be kind of difficult to receive strong signals. The Soviet magazine Raddio claims that strong signals, 5, 7, 9, you know, the range; not 19.995 megacycles, or the other 24 in that range, were received. In our parlance, it would be 5 over 5, or 5 over 4, in terms of intensity. In other words, highly readable and very loud signals were received by hams when the vehicle was 280,000 miles out from the earth.

Also you mentioned that because the vehicle was below the horizon it wouldn't be heard on this side of the world until the vehicle was out about 60,000 miles.

On the other hand, isn't it true that in the 20-megacycle range, due to ionospheric refraction, is kept within the confines of the earth's atmosphere and just can't, you know, the old skip-distance effect, wouldn't it be possible that these signals, if they were being heard very loud and very clear in the Soviet Union, could also be heard in the rest of the free world?

Mr. SLAVIN. You must rule out the skip-distance because you are already above the ionosphere.

Mr. MALLAN. That is true; you are above it. But wouldn't there still be refraction?

Mr. SLAVIN. Very little.

Mr. MALLAN. Then why would the Soviet hams be able to hear these signals so loud and clear?

Mr. SLAVIN. I would point out that the signals—if I may answer this, Mr. Chairman——

Mr. MALLAN. Sure.

Mr. SLAVIN (continuing). That the signals which were received by some of the stations in the United States were received well up to the time of closest approach to the moon of this vehicle. Some strong, some weak, probably depending upon conditions and on the sensitivity of the observer's receiving equipment.

Mr. ANFUSO. I think that is all, Mr. Mallan.

Mr. MALLAN. Just this one more very important question——

Mr. ANFUSO. I don't want to create the wrong impression here that one of the finest branches that we have in our Government, such as the Air Force, is on trial.

Mr. MALLAN. No, no; believe me, I have the greatest respect for the Air Force, and anyone in the Air Force will tell you that.

Mr. ANFUSO. All right, just a final question, because we must move on.

Mr. MALLAN. The Air Force is my favorite service, believe me.

Mr. ANFUSO. It is your favorite service?

Mr. MALLAN. Yes, it is. All my writing is about air research and development.

Mr. ANFUSO. One final question.

Mr. MALLAN. The final question is—and this was suggested to me by Dr. John D. Krause, director of the Ohio State Radio Observatory, who does not agree that the Goldstone signals are any definite scientific proof of a vehicle out in space, he suggested that motions—I will let you see the diagram, it is very simplified—but a single television transmitter on channel 8 or even an omnirange beacon signal—and those are all omnirange high-frequency stations surrounding Goldstone. And I might mention that Goldstone had never before nor never since observed or listened on 183.6—that Dr. Krause stated that these signals and the motions of them, the apparent motions and the widening angle, could have been the result of a series of aircraft moving across the airways, and that is a very busy air traffic area, and passed a certain point in either direction reflecting signals at intervals of an hour apart or so.

This might, because of the rotation and because of the movement of the moon itself, give the impression that the angle was widening the signal source in relationship to the moon.

Could you comment on that?

Dr. CURTIS. I would prefer to pass that question along to the Goldstone people, rather than my trying to answer for what their equipment does.

Mr. MALLAN. Would this be possible? That is all I was asking.

Dr. CURTIS. Highly improbable.

Mr. ANFUSO. That is all.

Mr. MALLAN. Thank you very much.

Mr. ANFUSO. Thank you very much, Dr. Curtis, and Mr. Slavin. You will stand by for a little while until we hear Dr. Stewart.

Dr. Stewart, would you please state your full name and the position you hold for the record?

STATEMENT OF DR. HOMER JOSEPH STEWART, DIRECTOR OF THE OFFICE OF PROGRAM PLANNING AND EVALUATION, NATIONAL AERONAUTICS AND SPACE ADMINISTRATION

Dr. STEWART. My name is Homer Joseph Stewart. I am Director of the Office of Program Planning and Evaluation at the National Aeronautics and Space Administration.

Mr. ANFUSO. Do you have a prepared statement, Dr. Stewart?

Dr. STEWART. I do not, Mr. Anfuso. I discussed this problem with you in January, if you recall. I am here to respond to any questions you may wish to ask.

Mr. ANFUSO. Supposing I ask you some questions, then, in open session, and then we will reserve some for the closed session.

Dr. STEWART. Very good.

Mr. ANFUSO. What do you mean when you say—and you stated this—that the Goldstone station was tracking lunik with automatic tracking equipment?

Dr. STEWART. The Goldstone antenna is a radio direction-finding device which tracks in two angles, or which is movable in two angles, I should say. The process of moving it to follow a signal is referred to as tracking. Tracking is also used in a slightly different sense with radar, where you also follow it by a range measurement.

In this case the two angles are all that are measured. If the signal is of good enough quality you can feed the input signal into a computer to drive the antenna so the automatic follows it. If the signal is of a poor quality, you may have to manually assist or you may have to do it purely manually.

Mr. ANFUSO. Would you please also explain what you mean by "cross-bearings," "ranges," and "bearings"?

Dr. STEWART. Bearing is a word that is used for an angle measurement. It is of nautical background. I don't recall where I used the words "cross-ranges".

Mr. ANFUSO. "Cross-bearings."

Dr. STEWART. This just means more than one angle measurement.

Mr. ANFUSO. Did any other station other than Goldstone monitor 183.6 mc?

Dr. STEWART. I believe there were some other attempts with ordinary receiving equipment. So far as I know, there was no proper direction-finding equipment used on this frequency.

Mr. ANFUSO. What about Jodrell Bank?

Dr. STEWART. Yes; I had overlooked that. I am sorry. I should apologize to Mr. Lovell.

Mr. ANFUSO. That is right. Did this station receive any signals on that frequency?

Dr. STEWART. Jodrell Bank? So far as I know, they did not.

Mr. ANFUSO. Can you tell us why not?

Dr. STEWART. I think it would be best to discuss this in executive session.

Mr. ANFUSO. Fine. What guidance accuracy is indicated by a 4,700-mile closest approach to the moon?

Dr. STEWART. That is a question that you can't answer in an unequivocal manner. You can make assumptions as to what the intent of the flight was, and from that you can deduce an asnwer. The answer is only as good as your initial assumption, however.

For example, if they were trying to make a miss of 4,000 miles and they made a miss of 4,000 miles, this would be very good, indeed. If they were trying to hit the moon in the center and missed by 4,000 miles, then there is a significant but still not large guidance error involved.

As I recall, in the discussion with you in January, I had made a very crude estimate as to what this would imply. At the time I estimated this would imply a velocity error of the order of 10 feet per second.

Mr. ANFUSO. Would better guidance be required for a moon shot made from the U.S.S.R. than from one made from the United States?

Dr. STEWART. I think the guidance problem is essentially identical in the two places. The particular trajectories look somewhat different because they are launching from a higher latitude, and the propulsion problems are very slightly more difficult, but not substantially.

Mr. ANFUSO. There has been a lot of talk here about the size of the sputniks. Could reflectors from a sputnik increase the apparent size of the sputnik?

Dr. STEWART. I don't think reflectors could. However, as you know, we have been interested in the use of balloons to give large size objects which really are really still quite light, as a means of carrying out certain kinds of experimentation.

Mr. ANFUSO. Where would the balloons be?

Dr. STEWART. The balloon would be the object you would look at.

Mr. ANFUSO. Is there a possibility that the Russians could have used reflectors right inside the sputnik so as to magnify the size of it?

Dr. STEWART. I can't conceive of any way of doing that.

Mr. ANFUSO. Any other questions for now, before we get into closed session?

No questions on the other side?

All right, I think this meeting will now go into closed session. You stand by, Dr. Stewart.

(Whereupon, at 11:20 a.m., the committee proceeded into executive session.)

www.ingramcontent.com/pod-product-compliance
Lightning Source LLC
Chambersburg PA
CBHW050412110426
42812CB00006BA/1869